Realms of the Sea

Realms of the Sea

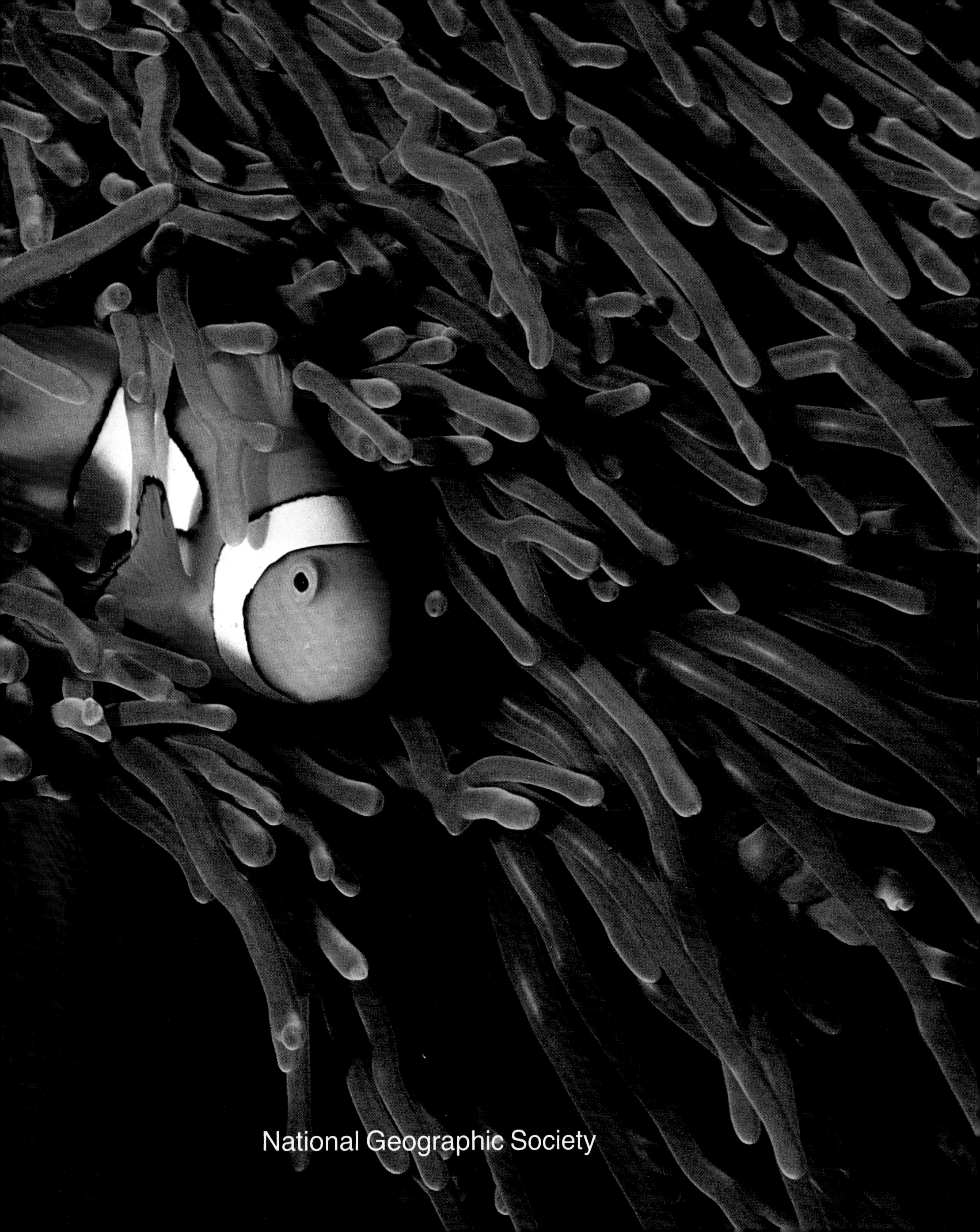

National Geographic Society

Realms of the Sea

By Kenneth Brower

Published by
The National Geographic Society

Gilbert M. Grosvenor
*President &
Chairman of the Board*

Michela A. English
Senior Vice President

Robert L. Breeden
*Executive Adviser to
the President for Publications
& Educational Media*

Prepared by
National Geographic
Book Division

William R. Gray
Director

Margery G. Dunn
Senior Editor

Charles O. Hyman
*National Geographic
Book Service*

Staff for this book

Leah Bendavid-Val
*Project Editor &
Illustrations Editor*

Elizabeth L. Newhouse
Text Editor

David M. Seager
Art Director

Joyce B. Marshall
Chief Researcher

Mary B. Dickinson
Editor-Writer

Leah Bendavid-Val
Jeffrey P. Cohn
Mary B. Dickinson
Elizabeth L. Newhouse
Portfolio & Caption Writers

Cathryn P. Buchanan
Paulette L. Claus
James B. Enzinna
Lise Swinson Sajewski
Anne E. Withers
Editorial Researchers

David Ross
Illustrations Researcher

Michael S. Frost
Laurie A. Smith
Artemis S. Lampathakis
Illustrations Assistants

Charlotte Golin
Design Assistant

Karen F. Edwards
Traffic Manager

R. Gary Colbert
Executive Assistant

Sandra F. Lotterman
Teresita Cóquia Sison
Marilyn J. Williams
Editorial Assistants

George I. Burneston, III
Indexer

Richard S. Wain
Production Manager

Emily F. Gwynn
Production Assistant

George V. White
John T. Dunn
*Manufacturing &
Quality Management*

Margaret Sedeen
Contributor

David Doubilet
Emory Kristof
Bill Curtsinger
Flip Nicklin
Photographs

Michael Hampshire
Paintings

First edition: 111,000 copies
278 pages, 188 illustrations.

CONTENTS

Foreword

by Robert D. Ballard

arth is unique among the planets because of the great seas that blanket 71 percent of its surface. Were it not for the water contained in the ocean, earth would be intolerably hot by day and freezing by night. Our ocean absorbs and stores the sun's energy, moves it around the globe, and releases it back into the atmosphere, thus tempering our climate. The ocean's energy and the enormous diversity of elements dissolved in its waters gave rise to life about three and a half billion years ago. Today, the seas teem with multitudes of creatures comprising hundreds of thousands of species.

Though we've always been curious about the nature of the seas, lack of access to the depths kept us largely ignorant of even the most prominent ocean floor features until very recently. The 46,000-mile Mid-Ocean Ridge, which snakes through the major ocean basins like the seam of a baseball, was first seen only in 1973. The next year, I and other team members of Project FAMOUS (French-American Mid-Ocean Undersea Study) discovered the ridge's newly emergent volcanoes spewing lava that oozed across the ocean floor like toothpaste from countless tubes.

In 1977, along a Pacific stretch of the Mid-Ocean Ridge called the Galápagos Rift, we found a series of underwater springs, hot water escaping from vents in the fresh lava flows. And to our surprise, around the springs we saw amazing concentrations of strange life-forms—tube worms five feet long, with red-tipped heads that, when cut, bled humanlike blood; huge white clams more than a foot long; mussel shoals; pink fish; and a host of other animals. How could such great concentrations of life survive in a world of total darkness, where photosynthesis is not possible? The answer lies in the hot springs. Discovery of the chemical reactions that take place within them revolutionized our thinking about the ocean, its chemistry, and the origins of life on our planet.

But we have only begun to explore the depths of the sea. We know more about the surface features of Mars than we do about the ocean bottom. To date, we have seen less than one percent of the deep seafloor.

A diver explores an underwater cave in the Russell Islands group of the Solomon Islands.

We began using deep-diving submersibles in this century. The first such crafts were called bathyscaphes, the deep-sea cousins of balloons. The idea was simple: A pressurized capsule carrying a passenger was hung beneath a giant, gasoline-filled steel balloon. When the passenger wanted to go down, he released gasoline from the balloon. When he wanted to go up, he dropped steel shot from a magnetized drum. My two journeys to the ocean floor in bathyscaphes left lasting impressions. The first, in 1973 aboard *Archimède,* ended in a fire at 8,500 feet. The second, in 1977, was in *Trieste II.* That dive ended when we crashed into the side of a volcano 20,100 feet down.

We replaced the early diving vehicles with modern manned submersibles like *Alvin,* built in the 1960s at the Woods Hole Oceanographic Institution. With *Alvin,* I have gathered ocean floor data in hundreds of dives, but the work is slow. Once you are inside the submersible, it takes two and a half hours or so to travel to the ocean bottom in the morning (an average distance of 12,000 feet), and another two and a half hours to return to the surface at day's end. That leaves only about four hours for actual exploration.

That lack of efficiency has been remedied with our newest creation, a remotely operated vehicle, or ROV, we at Woods Hole call *Jason.* It weighs well over a ton and can dive to 20,000 feet, which means it should be able to reach 98 percent of the ocean floor. Once *Jason* is in the water, it can stay there, working 24 hours a day, for days or weeks at a time without returning to the surface.

Paradoxically, the more we learn about the ocean, the more we seem to abuse it. Every year, millions of tons of wastes from homes and factories and millions of barrels of petroleum from oil spills enter the ocean. All kinds of plastic packaging, including bags and six-pack rings, entangle and strangle sea turtles, seabirds, sea lions, and fish.

Not only do we dump poisons in, we take too much life out. Unfurling its drift nets in 1989, a fleet of fishing vessels out after squid accidentally snared 58,100 blue sharks, 914 dolphins, 141 porpoises, 52 fur seals, 25 puffins, and 22 marine turtles. Drift-netting for tuna and squid kills 800,000 seabirds and up to 120,000 dolphins, whales, and seals each year.

Alarmed conservationists and an increasingly informed public have been pressing for laws to help prevent further devastation. Knowledge may be leading to actions that will save earth's underwater wildernesses.

This book, with its exquisite photography and graceful and informative text, aims to heighten the reader's appreciation of the ocean's wildness and mystery. I hope that through these pages the reader will sense the rhythm of the sea and hear its music, and understand that our life-filled, life-giving ocean is among our greatest treasures.

Otis Barton, designer of the bathysphere, prepares to celebrate its 50th anniversary with a dive in the Johnson Sea Link, a high-tech submersible.

9

The Ocean World

If there is magic on this planet, it is contained in water.
Loren Eiseley

Water, the magic element, is the one substance on earth that exists naturally in three states: solid, liquid, and gas. It has the greatest heat capacity of any liquid occurring naturally on the planetary surface, with the exception of ammonia. Its freezing, melting, and boiling points are aberrantly high for a molecule of its light weight. Of all liquids save mercury—a magical fluid in its own right—water has the highest surface tension. Water has a preternatural affinity for other substances and an unmatched ability to dissolve them. *Water is less dense as a solid than as a liquid,* and in this runs counter to most of creation. Water is the medium in which that other piece of magic—life—had its origins, and in which the chemistry sustaining life takes place.

Almost all the water on earth—97 percent or 320 million cubic miles of it—is in salt ocean. If the magic on this planet is contained in water, then most of that magic lies in the seas.

There is water elsewhere in the universe—in the dirty ice of Halley's Comet, for example, but nowhere on Halley's, or on any other comet, or anywhere else we know, does water pool deep and cool and blue. There may be *oceans* elsewhere. Titan, Saturn's giant moon, seems to have an ocean of ethane perhaps half a mile deep. Titan's may be a beautiful ocean. Sunsets over it may be lovely, as a weak and distant sun winks out in Titan's nitrogen atmosphere. But the Titanic ocean is not the sort of place where the life we know could endure.

There once were oceans and rivers of water on Mars, photographic evidence of old canyons and washes suggests. These have long since evaporated. Mars is now a desert planet, with a little ice of carbon dioxide at the poles. The only known water planet is our own. No other watery blue-green orbs circle this sun. For life-forms based on carbon, this third sphere is the only known oasis.

Ocean is what distinguishes us. Its breath swirls and marbles the planetary surface in the ceaselessly changing patterns of the clouds. Its capacity for heat controls planetary weather, stabilizing earth's surface

Early morning clouds drift over the coral-fringed Duke of York Islands in the Bismarck Sea. These warm equatorial waters lie at the heart of the coral reef belt that spans the tropics. Temperature and light determine the survival of the creatures of the reef and, in varying degree, affect all life in the oceans.

Pages 12–13: A cascade of silversides confronts a diver investigating Devil's Grotto in a Caribbean reef off the Cayman Islands. Enormous schools of these fish gather here in summer, possibly to spawn.

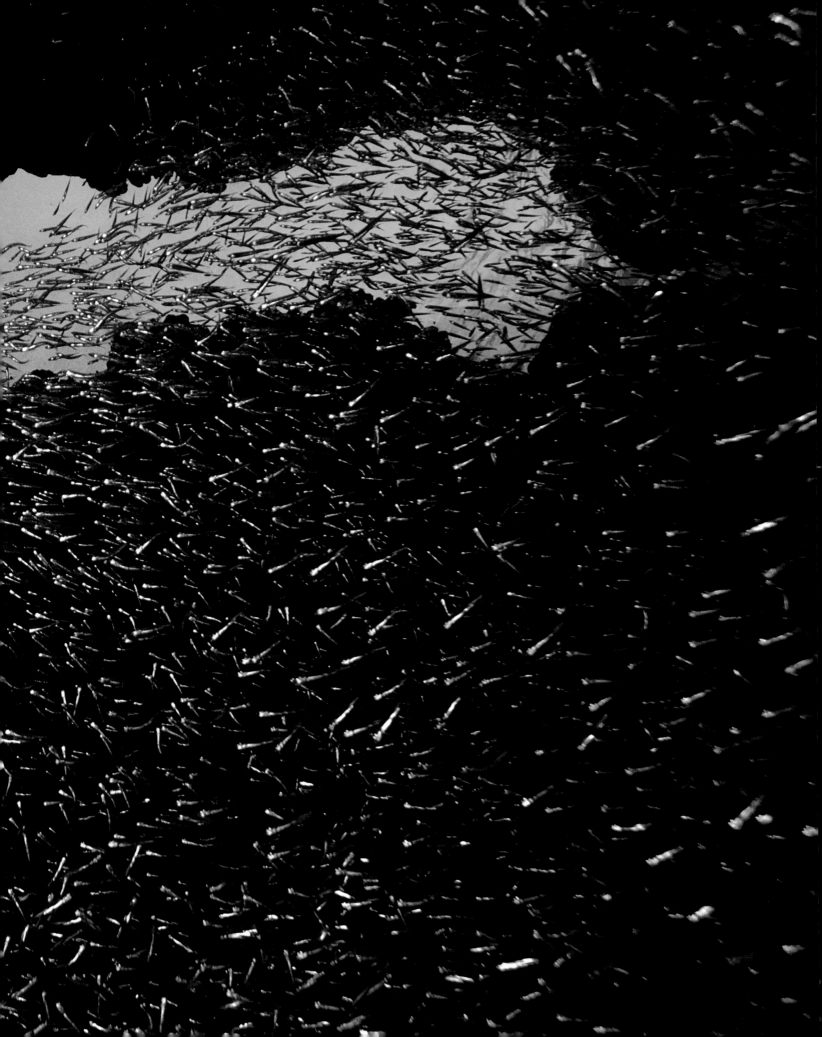

temperatures and insulating life in the ocean. It is where the planet lives, a habitat with vastly more volume for living than land.

Earth's oceans are interconnected, all one. Ocean makes a single reality, but one with a thousand moods, a thousand faces. It howls cold and mountainous in the circumglobal millrace of the roaring forties; it calmly mirrors the clouds in the doldrums along the Equator. It rises warm twice daily up the aerial roots of Indonesia's mangroves; it cracks like rifle fire in Arctic ice. It shades green along the margins of the continents, where upwellings and an influx of nutrients support blooms of phytoplankton; blue in the marine deserts of the open ocean; turquoise over the white-sand floors of atoll lagoons. It is full of surprise and paradox. Along the barrier reefs of the tropics, the blue monotony of the open ocean ends suddenly in the wildest colors under the sun, and the sterility of open ocean gives way to the community of corals, one of the richest ecosystems we know. On the frigid floors of the sea, there are oases of life and heat. In the total blackness of the lower depths, there are the convivial lights of bioluminescent organisms.

The magic in water is wonderfully simple. It's more than that; it's elemental. In a water molecule, two atoms of hydrogen—the lightest, commonest element in the universe—join one atom of oxygen, the commonest element in the planetary crust. The two hydrogen atoms form a V-shaped molecule with the single oxygen atom, producing an electrical asymmetry. A water molecule is dipolar. It has a slight positive charge at the hydrogen end, a slight negative charge at the oxygen end. From that dipolarity flows most of the behavior of the ocean and a good deal of the existence we know.

The dipolarity of water molecules causes them to bond to one another. The positive charge at the hydrogen end of one molecule attracts the negative oxygen end of the next. This attraction—the hydrogen bond—is what holds the seas together.

At the air-water boundary, the hydrogen bond creates surface tension. Surface tension is the force that gathers water into droplets of spray. Surface tension interacts with wind to start the ripples that grow to waves. It forms a "skin" that allows objects heavier than water to float on top. It is the reason that water-striding insects like *Halobates* can skate on the surface of the sea.

The hydrogen bond influences the viscosity of water. Viscosity is a measure of the force necessary to separate the molecules of a liquid and allow passage through. Viscosity simultaneously buoys sea creatures and obstructs them. In small planktonic animals whose mission is simply to hold position at a certain depth, viscosity breeds excess. The plankton tend to be all attenuated, feathered, and filigreed to increase their surface area. In elaborating themselves this way, they work *with* viscosity, instead of against it, and slow their rate of sinking.

Viscosity is temperature-dependent, decreasing as water temperature increases. Because warm waters are less viscous, tropical plankton are more extravagantly ornamented than the plankton of colder latitudes.

In large, mobile creatures whose mission is ahead, the struggle is *against* viscosity and density, and that struggle trims away all excess. A fish moving through water generates turbulence, which increases frictional drag. The various shapes of fish are various solutions to the problems of viscosity, density, and drag. The cut of the fins—tail fin forked in the jack, rounded in the grouper, lunate in the marlin—is in large part for slicing the hydrogen bond. The great scythe of the thresher shark, the siphon of the squid, the wings of the manta, are all to achieve velocity against viscosity, distance against density. The hydrogen bond, or the need to overcome it, slipperied the salmon's sides, sleekened the flanks of the dolphins. The odd asymmetry of the water molecule makes for the lovely symmetry in the flukes of whales.

The heat capacity of water is high, thanks again to the hydrogen bond. It takes large amounts of energy to overcome the bond and change water from liquid to gas. The temperature at which water vaporizes and its boiling point are high. That boiling point—212°F—occurs naturally on the surface of the earth only at sites of geothermal activity such as volcanoes and hot springs. Water's high boiling point, together with the planet's relative geothermal tranquillity, its safe distance from the sun, and the buffer of the atmosphere, keeps the earth's seas from evaporating away. Water's high heat capacity makes the seas the most important moderator of earth's climate.

The oddest feature of water, perhaps, is its temperature-density relationship. Most liquids become denser as they are cooled. For fresh water this is true to a point—about 39°F—at which the trend reverses and the water becomes less dense. When fresh water is chilled to this temperature, the hydrogen bond works more of its magic, forming water molecules into six-sided ice crystals, which take up more space than liquid water molecules. With fewer molecules in a given volume, solid water is less dense than, and floats on top of, liquid water; ice in forming rises. While a pond or lake of nearly any other fluid would freeze from the bottom up, a body of fresh water freezes from the top down.

Oceans generally freeze from the top down too, but there is a difference. The salts in seawater lower its freezing temperature. When seawater does freeze, ice crystals of fresh water form at the surface, leaving salts behind in the surrounding water. This cold, salt-rich water becomes

15

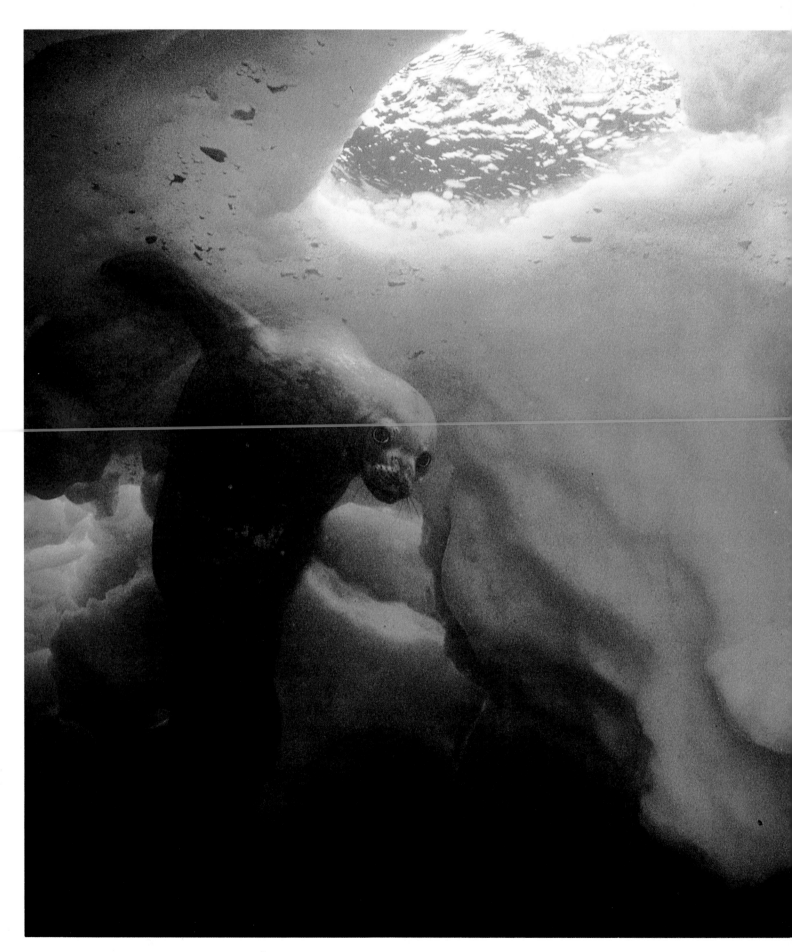

dense enough to sink and is replaced by deeper water. Because of its low freezing temperature, deep seawater generally freezes only in certain places at the Poles. If water behaved like other liquids, our polar seas would be solid ice.

Water, thanks again to the hydrogen bond, is the nearly universal solvent. For billions of years, rains have fallen on the mainlands; rivers have flowed down to the sea, carrying compounds in solution; the seas themselves have been at work washing away headlands, dissolving the minerals of their beds. Today the dissolved minerals we call salts compose about 3.5 percent of seawater. The ocean contains 160 million tons of salt per cubic mile.

The magical simplicity of water, in this matter of salinity, gives way to dizzying complexity. Seawater is a very complicated chemical substance. "If sea water were found only in some small, remote lake, chemists and biologists would doubtless study it intensively as a most precious liquid," oceanographer Robert Miller has written. The sea carries, in solution, traces of a major part of dry creation. We speak of soluble and insoluble materials, but everything, in truth, is soluble—some things just more rapidly so than others. Of the 103 known elements occurring in nature, most have been discovered in seawater, and it is likely that the rest will be detected there someday.

Seas are saltiest where they are very hot or very cold. In evaporating, as in freezing, water leaves its salts behind. In enclosed seas where the sun beats down—the Dead Sea, the Red Sea—surface evaporation concentrates salts. In polar seas, as we have seen, ice formation accomplishes the same thing. Because salt depresses its freezing point, seawater in the deep polar seas and deep ocean can grow colder than 32°F, the freezing point of fresh water, yet remain fluid.

It is the temperature-density-salinity relationship that has arranged the multistoried architecture of the ocean. The seas are stratified. The top few hundred feet form a mixed layer, roiled by the wind, in which phytoplankton bloom and almost all the organic matter in the oceans is produced through photosynthesis. Between 600 and 1,000 feet, in most of the ocean, the mixed layer ends and the layer called the thermocline begins. In the thermocline, temperature falls sharply to within a few degrees of 32°F. Below the thermocline, at about 3,300 feet, temperature, density, and salinity change very little all the way to the bottom. This is the deep ocean, by far the sea's thickest layer. In it no primary production occurs, except around hydrothermal vents. There is no flora, just fauna—parasites and predators of one sort or another.

It is temperature-density-salinity that drives the ocean's deep currents. As water grows colder, denser, and more *(Continued on page 25)*

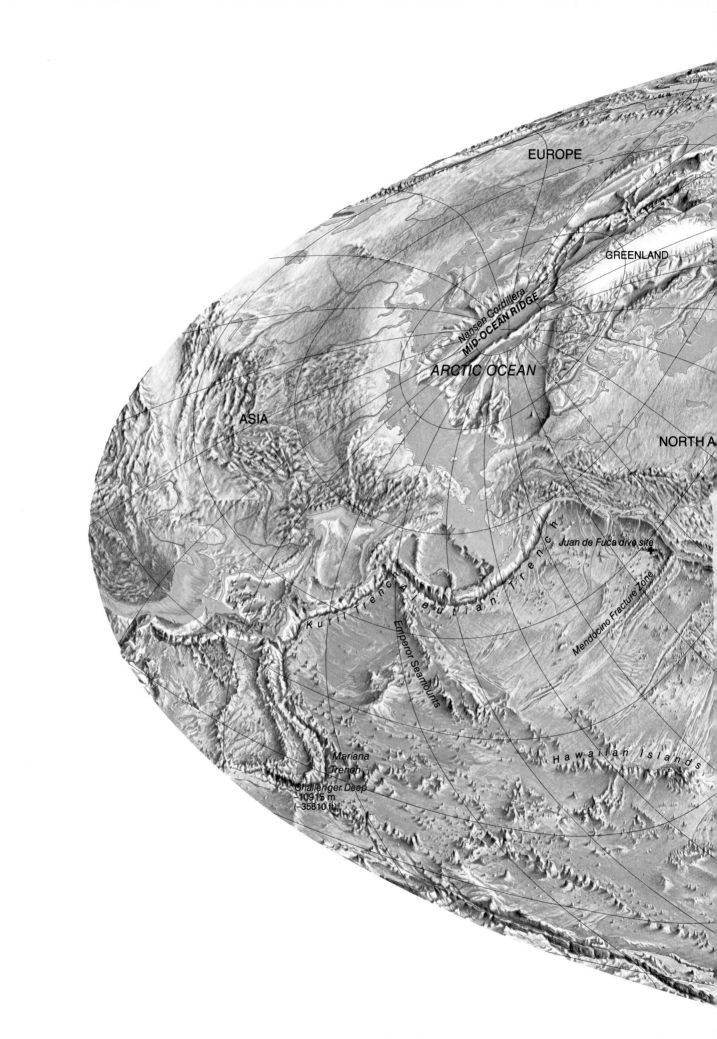

EUROPE

GREENLAND

Nansen Cordillera
MID-OCEAN RIDGE

ARCTIC OCEAN

ASIA

NORTH A

Juan de Fuca dive site

Mendocino Fracture Zone

Aleutian Trench

Kuril Trench

Emperor Seamounts

Hawaiian Islands

Mariana
Trench

Challenger Deep
−10915 m
(−35810 ft)

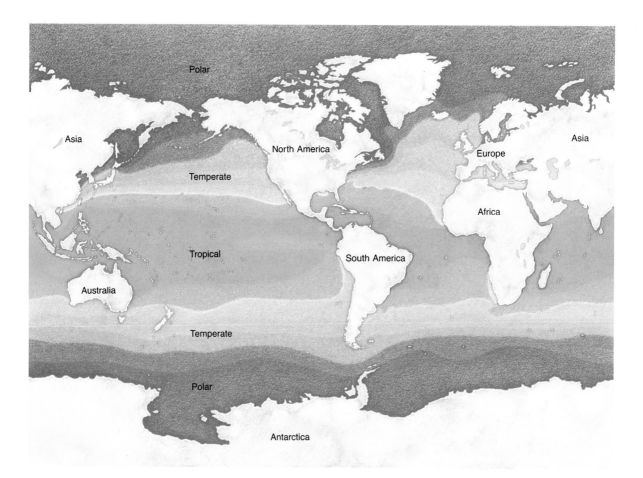

Polar

Asia

North America

Temperate

Asia

Europe

Africa

Tropical

South America

Australia

Temperate

Polar

Antarctica

Home to a huge variety of life-forms, the oceans cover more than 70 percent of the globe's surface. Because seas flow in a continuous stream around the continents, ocean species can spread widely. On land, temperature is only one of the factors that determine where plants and animals live; in the oceans it is the overriding one.

The map above illustrates the oceans' temperature zones. Polar regions, north and south, merge into two temperate zones separated from each other by warm tropical water. Ocean temperatures range from 27°F at the Poles to 86°F in the tropics. Only the temperate zones show a marked seasonal range.

These temperature divisions apply only to surface waters; they have no impact on the ocean deeps. Temperatures drop on the way down to the ocean floor (overleaf).

On these abyssal plains, fractured by trenches and dominated by the 46,000-mile-long Mid-Ocean Ridge that seams the earth, life faces some of its greatest challenges and takes on strange and wonderful forms.

Pages 22–23: The power of the oceans gathers in the curl of a wave as it crashes on shore. Wind, wave, and current fuel the dynamics of the seas.

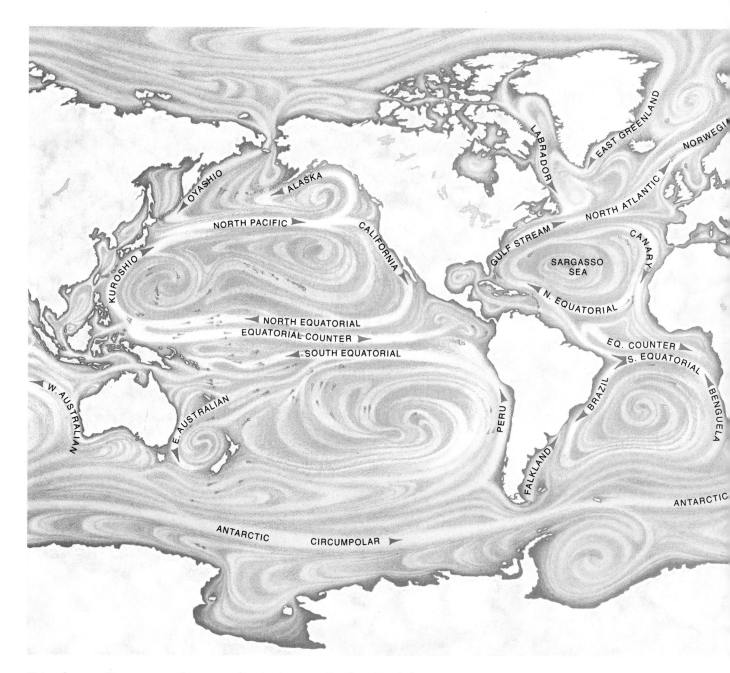

Driven by prevailing winds, the ocean surface swirls in perpetual motion. Winds push it along in great streams, or currents. Deflected by continents, the currents flow in circular patterns. Because of earth's rotation, major currents veer clockwise in the Northern Hemisphere, counterclockwise in the Southern Hemisphere—a result of the Coriolis effect.

The sinking of cold, dense, saline water drives deep currents. They interact with surface currents to circulate the oceans; oxygen is carried to the deeps by the sinking water, and life-giving nutrients are brought to the surface in cold-water upwellings.

N. EQUATORIAL
EQ. COUNTER
S. EQUATORIAL
AGULHAS
CIRCUMPOLAR

Pages 26–27: A humpback whale breaches off Alaska. Ranging in size up to the blue whale, largest animal in the world, whales are found in all the oceans.

saline, it sinks. Sinking, it displaces bottom water, which has to move. The coldest, densest, saltiest water in the ocean flows north from the Antarctic, moving along glacially, both in temperature and speed, hugging the bottom. A less dense Antarctic current rides north at a higher level. Between the two, an antipodal current crawls south from the Arctic.

In the upper ocean, currents are more complicated. They have origins outside the water molecule, causes as distant as the sun.

The middle latitudes of this planet receive more solar energy than the Poles. Uneven heating in the "ocean" of the air generates winds, and winds drive the currents in the watery ocean below. The great current systems of the earth—the Equatorial Current, the Equatorial Countercurrent, the Gulf Stream, the Kuroshio, the Antarctic Circumpolar Current, and the others—all begin in belts of wind. The currents are steered by the Coriolis effect. This effect, a consequence of the rotation of the earth, deflects currents to the right in the Northern Hemisphere, to the left in the Southern. It causes the major current systems of the planet to whirl clockwise above the Equator, counterclockwise below it.

If the ocean's great horizontal movements, the currents, have their origins in the wind, then so does the vertical movement of seawater. Upwelling is most pronounced along the western coasts of continents, where offshore winds blow surface water seaward, to be replaced by water from below. Cold-water upwellings, in their slow ascendancy, bring with them phosphates, nitrates, and other nutrients that have sunk into the dark. By returning those nutrients to the zone of sunlight where they nourish the growth of phytoplankton, upwelling has enormous influence on the fertility of the sea.

"Ocean, who is source of all," sang Homer.

Like many of the ancients, Homer was conceptually wrong about the shape of the ocean, but intuitively right about its primacy. For Homer's Greeks, as for the Egyptians and Chaldeans, earth was flat and ocean was an eternal river flowing around its rim. In the Egyptian scheme, a boat carrying the sun made a daily circumnavigation. This is not celestial mechanics or geography as we recognize them today, yet in some particulars the ancient views seem not so old. The primordial substance postulated by the Greek philosopher Anaximander—fluid, all-pervading, endowed with inherent powers of movement—sounds an awful lot like plain water. The Nu of the Egyptians—the primeval, chaotic sea in which the seeds of life floated—is not so very different from the Archeozoic seas described by modern paleontologists.

The ocean is an opaque, three-dimensional, storm-tossed medium unfriendly to human inquiry. Our *(Continued on page 30)*

25

Surface

EPIPELAGIC

Common Dolphin
(7 feet)

Flyingfish
(18 inches)

Blue Shark
(5 feet)

Ocean Sunfish
(3 feet)

Dall's Porpoise
(6 feet)

Pacific Barracuda
(3 feet)

Ray
(4 feet)

Phytoplankton &
Zooplankton

Swordfish
(12 feet)

Albacore
(3 feet)

Gray Whale
(50 feet)

650 feet

MESOPELAGIC

Snake Mackerel
(2 feet)

California
Smoothtongue
(3 inches)

Northern
Elephant Seal
(8 feet)

Lancetfish
(2 feet)

Lanternfish
(3 inches)

Oarfish
(23 feet)

Zooplankton

Scabbardfish
(4 feet)

Hatchetfish
(2 inches)

Sperm Whale
(to 60 feet)

Pacific Viperfish
(10 inches)

3,300 feet

BATHYPELAGIC

Fangtooth
(6 inches)

Pacific Sleeper Shark
(23 feet)

Grenadier
(2 feet)

Tripodfish

Deep-sea Eel
(2.5 feet)

Gulper Eel
(4 feet)

Sea Pen

Zooplankton

Brotula
(5 inches)

Brittle Stars

Deep-sea Sole

Sponges

Sea Cucumbers

Molluscs

Polychaete Worms

28

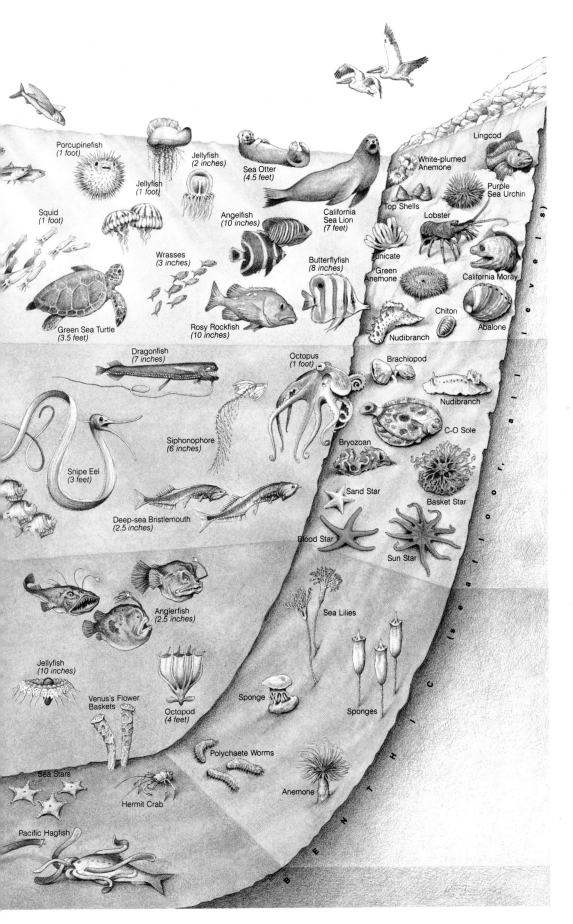

Porcupinefish
(1 foot)

Jellyfish
(2 inches)

Sea Otter
(4.5 feet)

Lingcod

White-plumed
Anemone

Jellyfish
(1 foot)

Squid
(1 foot)

Purple
Sea Urchin

Top Shells

Lobster

Angelfish
(10 inches)

California
Sea Lion
(7 feet)

Tunicate

Wrasses
(3 inches)

Butterflyfish
(8 inches)

Green
Anemone

California Moray

Chiton

Abalone

Green Sea Turtle
(3.5 feet)

Rosy Rockfish
(10 inches)

Nudibranch

Dragonfish
(7 inches)

Octopus
(1 foot)

Brachiopod

Nudibranch

Siphonophore
(6 inches)

C-O Sole

Bryozoan

Snipe Eel
(3 feet)

Sand Star

Basket Star

Deep-sea Bristlemouth
(2.5 inches)

Blood Star

Sun Star

Anglerfish
(2.5 inches)

Sea Lilies

Jellyfish
(10 inches)

Venus's Flower
Baskets

Octopod
(4 feet)

Sponge

Sponges

Polychaete Worms

Sea Stars

Anemone

Hermit Crab

Pacific Hagfish

Sea plants and animals
have adapted to condi-
tions in different horizontal life zones, but many
species swim or migrate
from one zone to another.
In this composite scene
from the Pacific Ocean basin, a sperm whale dives
down to hunt in the deeps.

Species of the open
ocean—known as pelag-
ic—are shown in exagger-
ated profusion; on the
right and bottom are ben-
thic species that dwell on
or near the seafloor. None
is drawn to scale, the tini-
est plankton being en-
larged as if seen through
a microscope.

Beneath the sunlit sur-
face waters, plants can no
longer exist. Light grows
dimmer until, at about
3,300 feet, it disappears al-
together. At all levels, sea
animals have developed
remarkably varied adapta-
tions for survival.

knowledge has advanced since the time of Anaximander, yet we have hardly scratched the surface. Our ignorance of the ocean is itself oceanic, as vast as the ocean basins, as deep as the abyss. The great oceanographer Sir Wyville Thomson, in his *Depths of the Sea,* recalls the fabulous ocean of his 19th-century childhood: "There was a curious popular notion, in which I well remember sharing when a boy, that, in going down, the sea-water became gradually under the pressure heavier and heavier, and that all the loose things in the sea floated at different levels, according to their specific weight: skeletons of men, anchors and shot and cannon, and last of all the broad gold pieces wrecked in the loss of many a galleon on the Spanish Main; the whole forming a kind of 'false bottom' to the ocean, beneath which there lay all the depth of clear still water, which was heavier than molten gold."

We laugh, but this curious popular notion persisted into our own century. Some people suggested that the *Titanic* might have come to rest on just such a false bottom.

In my own boyhood, I read and believed a book called *The Sea.* It was published in 1961, yet today seems as quaint as one of those medieval maps on which sea serpents loll and the wind puffs his cheeks. One photograph shows a diver and a hammerhead shark. "Few sights," the caption reads, "are more terrifying for a skin diver than the ugly form of an approaching hammerhead shark. The animal comes out of the murk like some prehistoric monster, zeroing in on its victim. Eyes at the tip of its grotesque head lobes enable it to see backward as well as forward. Using these fleshy projections for balance, it hangs motionless before it makes its attack. Unless the diver is a skilled spear hunter, . . . he may go down on a long list as another victim of the hammerhead, one of the most notorious man-killers of all the sharks."

Nearly everything about this is wrong. The form of the hammerhead is not ugly, but beautiful. The hammerhead does not zero in on victims, but approaches obliquely, in the cautious manner of sharks. Those fleshy projections are not fleshy, but cartilaginous. They are not used for balance; they enhance the shark's hydrodynamics, vision, smell, and electroreception. (The instrumentation in sharks is much more sophisticated than we dreamed 30 years ago.) The hammerhead does not hang motionless before it attacks. Hanging motionless is not an option for any shark. The list of human victims of hammerheads is in fact remarkably short. The hammerhead is dangerous to humans, but is nowhere near the most dangerous. Its rare attacks are involuntary manslaughter, a mistake, the shark spitting the human out in a kind of apology. The ocean, as we perceive it now, is a much less melodramatic place.

It is too much to hope that we have ended our penchant for error about the ocean. Some of our present theories will be corroborated, some laughed at, by future generations of marine scientists—who themselves, in time, will be proved both right and wrong. In recent decades, new technology has brought us new knowledge, particularly about the ocean floor and its life-forms. As we solve puzzles about the ocean, though, more will open up before us.

Sir Isaac Newton, in summarizing his life, chose a sea metaphor. "I do not know what I may appear to the world," he wrote a friend, "but to myself I seem to have been only like a boy playing on the seashore, and diverting myself in now and then finding a smoother pebble or prettier shell than ordinary, whilst the great ocean of truth lay all undiscovered before me."

For at least the next few centuries, any scientist gazing out beyond the breakers may find himself exactly where Newton did.

If the ocean holds most of the magic on this planet, it holds most of the mystery, too. At this moment, somewhere on Homer's wine-dark sea—somewhere on Newton's great ocean of undiscovered truth—a sperm whale bull has surfaced and is filling his lungs with a quick succession of breaths. His genus name is *Physeter*, Greek for "blower," and he is living up to it, repaying his oxygen debt after a dive. The mist of those breaths scurries away on the breeze. Downwind, a fishy smell passes for an instant, then nothing but the clean freshness of salt air.

Surface tension stands no chance with *Physeter*. The sperm whale grows to 60 feet and 50 tons. No creature—no thing—better illustrates how the magic simplicity of the water molecule can spin itself out into complexity. The whale in its composition betrays its origins—like the rest of us, it is more than 80 percent water, with blood as saline as the sea. But the whale is something else again. The magic of water, in the end, is just chemistry. The magic of the whale is alchemy. Somehow the salts of the sea organized themselves into *this*. Sixty feet and fifty tons of state-of-the-art biological gear. Largest brain in the solar system. Nose equipped with biosonar. Brow armed, perhaps, with a sonic artillery to kill or stun its prey. Head full of mysterious plumbing—great organ pipes whose purpose no one knows. The sperm whale cruises all latitudes. It swims the thermometer from one end to the other. It inhabits two universes—one of air and sunlight, the other of darkness, pressure, and cold—shuttling regularly between them. If the water planet has an ultimate production, it may well be *Physeter*.

The whale blows one last time. He dives. He shows his flukes. Into the great, wine-dark ocean of truth he disappears.

31

Deep Ocean

The flukes of the sperm whale, slipping under the glassy slope of the swell, leave a momentary slick behind. The wind ripples and erases it. Two hundred feet down, the whale passes through a school of yellowfin tuna, which flash silver and divide. The blue of the ocean grows darker, moodier, yet somehow bluer. Temperature drops. Light begins to fail, first the red end of the spectrum, then other wavelengths filtering out in a march toward the violet. Pressure builds. At sea level there are 15 pounds of pressure on each square inch of the body—one atmosphere of pressure. In the ocean, pressure increases by one atmosphere for each 33 feet of depth. The whale adds three or four atmospheres with each stroke of his flukes. As his lungs compress, the whale loses buoyancy. At a certain depth, he crosses a magic line, his positive buoyancy becoming negative, and a sudden new weightiness speeds his descent.

The whale enters the thermocline, a layer in which temperature drops precipitously and density rises. He plummets through the fast-gathering twilight and toward the bottom of the thermocline enters a narrow stratum called the SOFAR (Sound Fixing and Ranging) Channel. Here, temperature and pressure capture and concentrate sound, transmitting it great distances. The sperm whale hears the basso profundo of a humpback whale singing far away. The humpback bellows out a deep, slow tuba note, then repeats it with rising inflection, like some sort of query. Then it creaks like a castle door. Then it resumes the bass query; then suddenly, thrillingly, it shifts into high trumpet notes. It's all Greek to the sperm whale. His flukes drive him downward into an ever-tightening hydrostatic vise. At about 3,300 feet he leaves the thermocline, an inverted geyser of thermocline water marking his exit.

The steady beat of his flukes drives the whale down past the penetration of the last feeble photon from above. He sails down and down through total darkness. Colossal pressures distort his shape. He passes a few last mid-ocean fish—starbursts of bioluminescence as they flee, startled. His biosonar sends out series of clicks, and now and then acoustic

The aggressive fangtooth lives in a narrow, icy band of water below the range of sunlight. More than a hundred known species of fish have adapted to the harsh conditions of the oceans' largest realm—between 3,300 feet and near-bottom. That is one-fifth the number that inhabit the more hospitable twilight zone just above it.

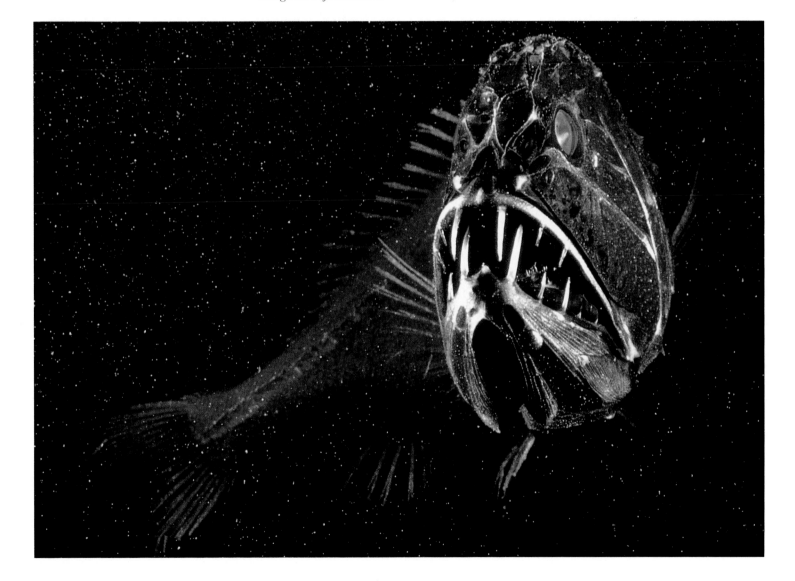

snapshots of anglerfish or dragonfish bounce back. The whale is not interested. It is after squid. In stygian cold and darkness a mile and a half down, it levels off and begins to search.

For their hunting grounds the sperm whales, greatest of toothed cetaceans, have chosen the vastest province of life on earth. The deep ocean is the biggest habitat on the planet. More than 60 percent of the planetary surface is covered by water deeper than a mile. That deepness very nearly makes a single worldwide system. All the deep oceans of earth, save the Arctic, are continuous; the deeps of the Pacific, Atlantic, and Indian Oceans all join the circumpolar deeps of the southern oceans.

For life on earth, our upper world of warmth, sunlight, and seasons is the exception. The dark, cold, dense, saline netherworld of the deep ocean is the rule. The vastest of habitats is sparsely inhabited. Flora is nonexistent, fauna impoverished. The deeps make a wet, profound, circumglobal desert. In the ocean, biomass decreases as depth increases. More than 80 percent of marine life, by weight, lives in the top 3,300 feet. The species living high, in the sea's sunlit layers where most biological productivity occurs, have first crack at that productivity. In the economy of the sea, trickle-down theory works as it does on land; the underclass lives low on the hog.

The sea is divided into two great realms, the pelagic and the benthic. The first term derives from the Greek *pelagos,* sea, and it describes all the ocean away from shores and bottoms. The second derives from the Greek *benthos,* deep sea, and describes the ocean floor. The two realms are further divided horizontally into zones. The uppermost zone of the benthic is an illuminated zone, which begins in the intertidal region and ends about 650 feet down. Below that are the zones of darkness: first the bathyal, from the Greek *bathys,* the deep; then the abyssal, from the Latin *abyssus,* bottomless gulf; and finally the hadal, from the French for "hell." It seems a scheme designed less by an oceanographer than by Dante.

By far the greatest portion of the ocean is pelagic. The uppermost layer is called the epipelagic, or euphotic (good light), zone. It begins at the surface and ends as deep as 650 feet, making a thin but vital epithelium. It is here that phytoplankton—the tiny plants responsible for all pelagic photosynthesis—do their work. Beneath this surface layer—from depths of 650 to 3,300 feet—is the mesopelagic, or disphotic (bad light), zone. Next, between 3,300 feet and the bottom, lie three layers comprising the aphotic (without light) zone.

The epipelagic zone is a clear, well-lighted place, a region of fish like the tuna, the kingfish, the sailfish, the marlin—fish of the sort engineers might have designed, working calmly and lucidly at drafting tables. These fish are the very latest in ichthyotechnology. The bodies of tuna are tapered, like aircraft fuselages, all fitted with caudal keels, finlets, and other ingenious hydrodynamic features. The marlin have heater organs to warm their brains. The cetaceans are more than modern; they are futuristic, with the largest brains to have evolved on this planet. The epipelagic is the zone where the ocean has its bright ideas.

Underneath, far more voluminous, is the mesopelagic zone of dimness and dark, inhabited by midwater squid and fish of habits weird and aspects strange. This is a transition zone. It is the floor of the zone of light, the ceiling of darkness. It is inhabited by creatures with exquisitely photosensitive eyes and bodies spangled with photophores. The number and diversity of inhabitants are greater than in any of the layers below, until one reaches the bottom.

Below and more voluminous still lies the true deep sea, the frigid blackness and spooky fauna of the aphotic zone. Here, the creatures look like bad dreams: dragonfish, viperfish, the listless or yawning melamphaid, blacksmelts, snaggletooths, daggertooths, fangtooths. "In these abyssal regions," naturalist William Beebe wrote, "there are fish which can outdragon . . . any mere figment of the imagination; crustaceans are there to which the gargoyles of Notre Dame, the fiends of Dante's Purgatory appear usual and normal."

We live in a relative universe, of course. The view of the deep as a gloomy Purgatory is mostly surface bias. It's the old surface-animal arrogance. The reduced skeletons and musculatures of deep-sea fish, which seem to us like degeneration, are actually a means of achieving neutral buoyancy. Those reductions, along with lowered metabolism, are evolutionary adaptations to the darkness and stillness of the depths. To a creature from the deep, heaven lies downward, and the stomach-turning near-vacuum of our surface pressures, the scalding temperatures, the cruel blinding light, the maelstrom of currents, the chaos of waves, the sickening swings in salinity, the frantic pace of metabolism and of life, make the purest sort of hell.

In the deep, the creatures are all scavengers or predators of one kind or another. The smallest scavengers are the copepods, amphipods, bacteria, and other tiny organisms that feed on the rain of particles from above. This slow-falling detritus is the "snow" that the early explorers of the deep saw through the windows of their bathyspheres and bathyscaphes. Small detritus particles settle at a rate of from 1 to 10 feet per day, taking 25 years to reach the floors of the trenches. It is manna, but manna of the slowest, sparsest sort. Occasionally, for the larger scavengers, more sizable gifts come down—a piece of a log, or a coconut husk,

or a fragmentary part of some large fish uneaten by predators above. A characteristic of many pelagic deep-sea fish is oversized jaws and undersized bodies. Because feeding opportunities are few and far between, the strategy is to eat as much as possible at each sitting.

Champions at this are one family of gulper eels. These gulpers are creatures out of a cartoon, part python and part pelican. They have giant jaws attached to slim, snakelike bodies that enable them, like many other deep-water fish, to swallow prey larger than themselves.

Some deep-water anglerfish can snap up prey as much as three times their length. Of deep-water pelagic fish, the anglerfish are the most diverse. They are sometimes smooth-skinned, sometimes spiny, generally of a velvety-black complexion, always abysmally ugly. Each female angler has the first spine of the dorsal fin modified into an illicium—a fishing pole with a luminous lure at the tip.

Female anglerfish generally mate with dwarf males. Male anglers of many species are tiny and parasitic. Finding a female, the dwarf male grips her with his little jaws, attaching anywhere—to her face, her flank, her belly, her back—the male is not particular. His mouth fuses to her skin, and eventually their circulatory systems merge. He ends his existence as hardly more than a sack of sperm.

For other examples of seeming degeneration—which, in fact, represent adaptation to the sparse conditions—witness the differences between the fish of the mesopelagic twilight and those of the dark. In mesopelagic fish, eyes are generally large to enormous, with retinas composed entirely of rods for detection of light. Photophores are often large and numerous, central nervous system fairly well developed, sense of smell moderately acute, skeleton well ossified, muscles well developed, swim bladder usually present, gills filamentous, heart and kidneys large. In fish of the deeper, dark zones, eyes are small (except in some male anglers before they are incorporated in their mates), photophores are small, central nervous system weakly developed, skeleton weakly ossified, muscles poorly developed, swim bladder absent or regressed, gills reduced, heart and kidneys small.

The deep-sea cephalopods—octopus and squid—are slower and less muscular than cephalopods of the upper ocean. Upper-ocean cephalopods are among the most advanced of invertebrates. Their nervous systems are so large that we dissect squid nerves to learn about our own. They have keen eyes and remarkable powers of form discrimination that enable them to detect different planes of polarized light. *Cirrothauma,* the deep-ocean octopus, in contrast, is gelatinous, floats like a jellyfish, and is blind.

The deep sea has its own physiology. Deep-sea fish and crustaceans

*Seafloor spreading contin-
ually adds new floor to the
oceans and disposes of old
seafloor in trenches.
Through a rift, or crack, in
the Mid-Ocean Ridge, mag-
ma wells up and solidifies
into new seafloor, which is
carried away in opposite
directions by the huge, rigid plates that make up
earth's shell (the litho-
sphere). When two plates
collide, the denser one will
subduct into a trench, often precipitating volca-
noes, as on the left side of
this painting. On the right
side, a continent is being
carried along on a plate.*

have rates of metabolism about 20 times lower than related organisms in
the upper ocean. The metabolisms of the sperm whale, the beaked
whales, and other deep-diving cetaceans slow considerably in diving,
though probably nothing like that. But the whales only work here, com-
muting down from the heaven of air and sunlight above.

The benthic environment—the seafloor—is richer than the pelagic
depths. Diversity and size in demersal fish, which are those living on or
near the bottom, are greater than in their deep-sea pelagic counterparts.
Yet the deep bottom remains a desert. As in the pelagic realm, diversity
and abundance drop with depth and as one moves outward from the
continents. Some benthic deserts, it is true, are more fertile than others.
They occur along the margins of the continents, where currents carry organic material down the continental slope from land. They can be found
offshore as well, under regions of high surface productivity, as in high latitudes and in tropical zones of upwelling. They are deserts just the same,

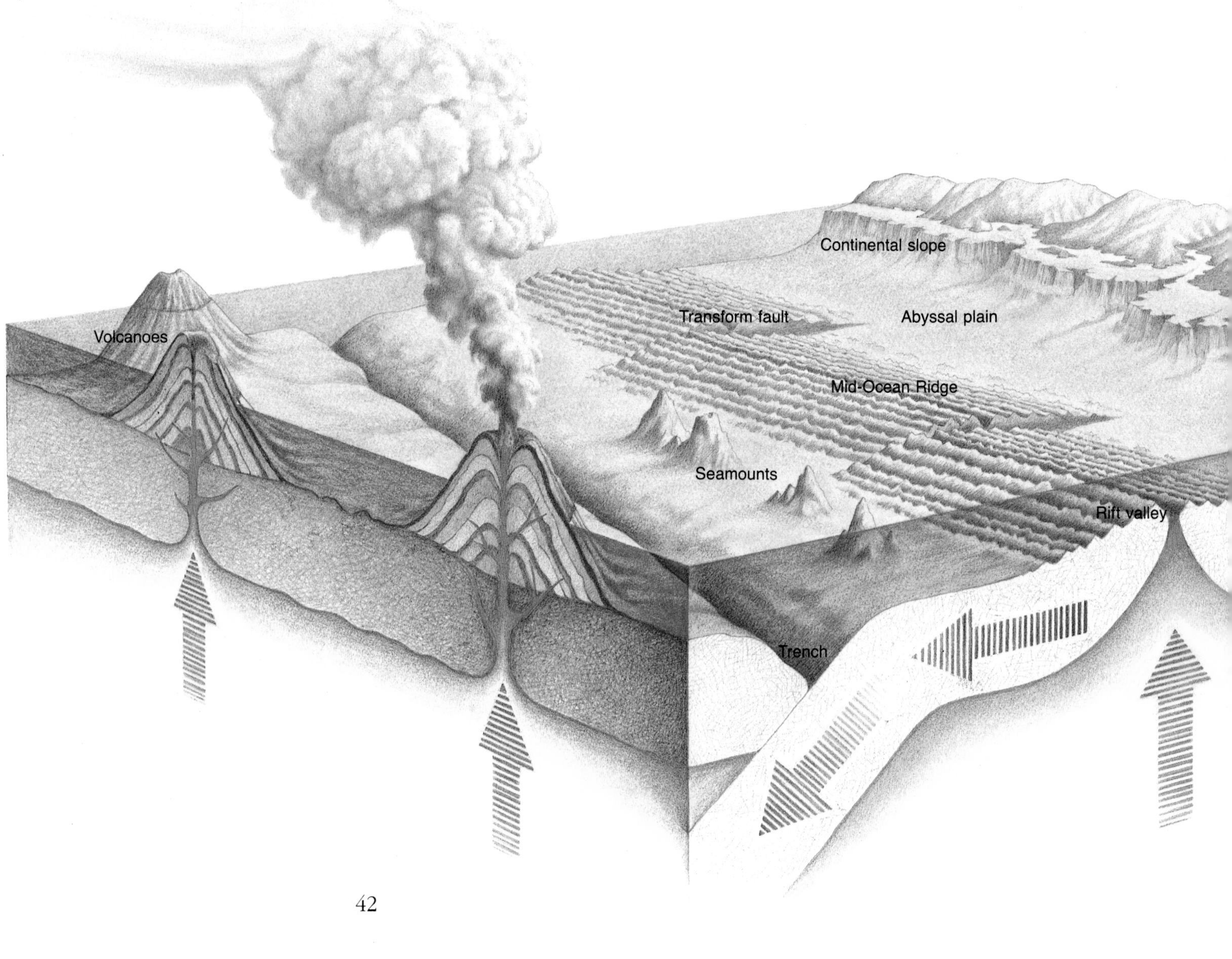

Volcanoes

Continental slope

Transform fault

Abyssal plain

Mid-Ocean Ridge

Seamounts

Rift valley

Trench

Pages 44–45: *An agile basket star snags zooplankton from the passing current at 200 feet. The creature's reedy arms branch and re-branch, and form knots around prey; periodically, multiknotted arms curl in-ward toward the mouth to deposit collected food. Remote-control submersibles and sophisticated underwater cameras have made possible photographs of such deep-sea creatures from beyond scuba diving range all the way to the ocean floor.*

the Mojaves or Sahels of the deep. The poorer deserts have just trace amounts of organic matter and are found, for the most part, under the big, mid-ocean surface gyres. They are the Saharas of the deep, and they look it—endless wastes of red-clay sediment, marked at great intervals by the track of a worm.

The themes of the pelagic deep repeat themselves along the bottom. There are the improbable forms—that of the tripodfish, for example, whose pectoral fins and tail fin have rays that are elongated and stiffened into stilts, on which the fish stands high above the abyssal mud. There is opportunism. The deep demersal fish are opportunistic from their very beginnings. As eggs, they are generally large and yolk-filled. They hatch into little adults. The austerities of the deep are such that no time can be wasted on childhood. There is the apparent degeneration. A number of benthic species in the deep are blind. The pogonophorans, or beard-worms, lack not only eyes; they lack mouth, digestive tract, and anus as well. The complex society of the hard corals disintegrates with depth. In the sunlight at the top of the ocean, hard corals are colonial and build the enormous, Byzantine, polychromatic miracles of symbiosis and bioengineering that we call coral reefs. In the deep ocean, hard corals occur only in solitary form, and below 10,000 feet they occur not at all. (At the latter depths, calcium saturation is so low that the calcium carbonate of the coral skeleton dissolves.)

Most invertebrate phyla have representatives well into the deepest zones, but there phyla begin dropping out of the fauna. They have had enough of scarcity and cold. As far as we know, there are almost no decapod crustaceans, few fish or sponges on the floors of the deep-sea trenches. Sea anemones have been found on the deepest floors. Anemones are perhaps the most adaptable of sea animals, a success from the rocky tide pools of the intertidal zone all the way down to the bottoms of the profoundest trenches. Gastropods—marine snails and their cousins—are found to the ultimate depths of the sea, though in the trenches shell formation is difficult because of the dissolution of calcium carbonate; the shells of snails are thin and delicate. Clams, polychaete worms, and spidery isopod crustaceans do well on the floors of the trenches. There is high endemism in these deep communities; 70 percent of the species of a given trench are found nowhere else.

"Roll on, thou deep and dark blue ocean—roll!" wrote Byron. "Ten thousand fleets sweep over thee in vain."

Insofar as they advanced mankind's knowledge of the deep, those fleets swept in vain indeed. For the first millennia of human navigation, we only skimmed the ocean. Ferdinand Magellan was the first, so far as

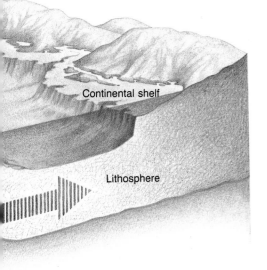

Continental shelf

Lithosphere

43

we know, to attempt a sounding of the open ocean. After negotiating the dangers of the strait that bears his name and reaching the calm ocean he called Pacific, Magellan lowered a sounding lead on all the line he had. It became part of the great navigator's legacy: the Strait of Magellan, the Magellanic Cloud, the Magellanic penguin; the circumnavigation of the globe, the naming of the Pacific, the demonstration that this greatest of oceans was more than 600 feet deep.

The most important scientific voyage in history, after Sir Charles Darwin's in H.M.S. *Beagle,* was that of H.M.S. *Challenger.* The *Challenger*'s odyssey, which began in 1872 and covered nearly 69,000 nautical miles in 42 months, first intimated to the world the reality of the deep. "Never," naturalist Sir William Herdman would say, "did an expedition cost so little and produce such momentous results for human knowledge."

Sir Wyville Thomson, leader of the *Challenger* expedition, had gone from a boyhood belief in a stratified sea—skeletons floating at one level, gold doubloons at another—to an adult conviction that below 300 fathoms life was impossible. The reasons were obvious to Thomson, as they were to most biologists of the time. Below 300 fathoms—1,800 feet—no light penetrated. At that depth, the water itself was viscous and dense, the hydrostatic pressures appalling. "At 2,000 fathoms," Thomson wrote, "a man would bear upon his body a weight equal to twenty locomotive engines, each with a long goods train loaded with pig iron."

In their calculations on the physics of the deep, Thomson and his contemporaries were mostly right. About the tenacity and adaptability of life on the water planet, they were entirely wrong, as the *Challenger* expedition demonstrated. The *Challenger*'s tow-nets brought a myriad of strange creatures from depths where life was thought to be impossible. The results filled 50 large volumes. The modern science of oceanography began on the decks of that little ship.

Well into this century, *Challenger*-era technology—dredges and thermometers—were the principal tools of deep-sea exploration. Then in 1930 came Otis Barton and William Beebe's bathysphere, a steel sphere with viewing ports lowered from the surface on a cable. In 1948, Swiss physicist Auguste Piccard built *F.N.R.S. 2,* the first bathyscaphe (from the Greek words for "deep" and "light boat"). In 1959, Piccard followed with *Trieste,* in which his son Jacques and Don Walsh of the U. S. Navy descended 35,800 feet, nearly 7 miles, into the depths of the Mariana Trench. If the bathysphere had a model, it was the cannonball. In the 60 years since Beebe's pioneer dives, that simple shape has rapidly evolved. The models for the newer generations of submersibles—*Deep Quest, Deep View, Archimède, Cyana, Alvin*—are the spaceship and the spider.

Wyville Thomson and other pioneer oceanographers of the 1800s

Cruising lights of the submersible Alvin illuminate a luxuriant hydrothermal vent community 8,000 feet down. Here, on a ridge in the eastern Pacific Ocean, warm water from deep inside earth percolates up through cracks in the porous rock, as cold, surrounding seawater seeps downward to replace it (red and blue arrows), and black "smokers" spew mineral-laden water superheated to 720°F. The heat and pressure inside earth's crust cause sulfate in the seeping seawater to convert into hydrogen sulfide, which on rising is used as fuel by the bacteria that form the basis of the community's food chain.

In this oasis, giant clams and tube worms predominate; grazing crabs scurry on rocks; ten-inch fish called eelpouts nestle among the clams or nibble at tube worms; occasionally an interloping grenadier fish looks in to feed. All this Alvin observes, but not too closely; the hot water could damage its windows.

destroyed the myth of the abyss as void. They disposed of the old notion of abyssal lifelessness, but they made no dent in the notion of abyssal changelessness. The oceanographers of the 20th century have been busy remedying that.

In the 1960s, a conceptual revolution, the theory of plate tectonics, shook up all the earth sciences. According to this theory—and few scientists doubt its validity—the planetary crust and upper mantle, or lithosphere, is divided into plates, on which the continents drift about, collide, separate, fragment, and meld. The seafloor is a key to the new concept. According to the theory, new seafloor is continually forming at "spreading centers," growing laterally as much as seven inches a year and building ocean ridges. The Mariana Trench and the other great trenches are subduction zones, where the floor dives back into earth's interior. The mantle's reingestion of seafloor generates volcanoes, which flank the trenches. The old static abyss has become a dynamic abyss. The deep sea is not just where things end up; it is where they begin.

We are revising our ideas about abyssal weather. Those glacially slow bottom currents of early-1970s textbooks have been rapidly accelerating. In 1979, instruments moored at the base of the Scotian Continental Rise, off New England, recorded a violent, week-long benthic storm. The western edges of the ocean basins are subject to such storms, it turns out. One-knot currents scour the bottom of sediment in some places, deposit great banks of it in others.

The most extraordinary discovery of 20th-century marine biology came in the late 1970s, between the Galápagos Islands and South America. The existence of hydrothermal vents in the deep sea had been postulated in the mid-sixties. In 1974, in the rift valley of the Mid-Atlantic Ridge east of the Azores, French and American scientists had observed what they took to be validation—a hydrothermal spring surrounded by metallic deposits precipitated from the hot water. In the Pacific, water temperature anomalies over the spreading center of the Galápagos Rift; temperature fluctuations in sediments near it; lines of mounds on those sediments; swarms of microearthquakes; a fish kill in which dead bottom fish floated to the top—all were evidence to marine geologists of vents in the new-formed oceanic crust of the rift. In May 1976, samples gathered by an unmanned submersible confirmed the existence of such vents.

Then in February 1977 the camera sled *ANGUS* was lowered to the vent area from the research vessel *Knorr*. Among the first 3,000 photographs it took were 13 showing thick beds of large clams and mussels at depths far deeper than such dense populations had any right to exist.

It was a biotic community entirely new and unexpected. Clustering thickly around the vents were some 25 families and subfamilies unknown

47

A robotic arm takes a sample from a black smoker on a hydrothermal vent off New Zealand, about 8,200 feet down. Shot from the Soviet submersible Mir *in the summer of 1990, this photograph is among the first to show creatures living on a deep-sea smoker.*

to science and 100 unknown species. If *Viking* had made such a find in some crevice on Mars, it would hardly have been more exciting.

Alvin followed *ANGUS.* In a 90-minute dive, the submersible reached bottom at 8,860 feet and skimmed across lava flows to the site. On that dive and subsequent trips, *Alvin* charted the infernal realm that its camera sled had discovered. In one area of venting, warm, 54°F springs emitted milky plumes of hydrogen sulfide. The springs were colonized by the big mussels and clams, for which the crew named them Clambake Vents. *Alvin* moved on to "Dandelion Patch," springs inhabited by odd, small, spherical animals. These new creatures suspended themselves from the rocks of the bottom on fine, weblike "legs" and looked like dandelions gone to seed. (The dandelions would prove to be siphonophores.) The sub lingered in what the crew called the Garden of Eden. The vents of the garden appeared to be the youngest of all. *Alvin* poked its mechanical arm and water-sampling probe into 60°F water rising from fissures in pillow lavas that looked fresh. The garden's outer margins were colonized by the dandelions, along with white crabs. Farther in were small anemones and small serpulid worms. Finally, at the heart of the community,

48

in the warm currents rising from the vents, were limpets, clams, mussels, and forests of long, white tube worms with furry, blood-red tentacles.

The vent communities flew in the face of all expectation. They turned dogma upside down. Biological production is supposed to be confined to the top of the sea—all food chains leading downward from photosynthetic plankton—yet here on the bottom, primary production was running wild. The explanation, it turned out, was that the hydrogen sulfide rising from the seafloor vents was—in a process called chemosynthesis—sustaining bacteria, the basis of the food chain.

Alvin's discovery brought theory full circle, in a curious way. Victorian scientists postulated that life began as "abyssal ooze." Early deep-dredging had brought up a gray material from the bottom, the scientists had preserved it in alcohol, named it *Bathybius,* surmised that it covered the ocean floor, and decided it was the "mother" of protoplasm and of the first organism. *Challenger* found some *Bathybius* too, of which a chemist named Buchanan analyzed a sample. He found no organic matter and finally demonstrated *Bathybius* to be nothing more than a precipitate of calcium sulfate resulting from the mixing of alcohol and seawater. In time, science came around to the conclusion that life had probably originated in the warm, sunlit shallows of some old sea. *Alvin*'s discovery did not revive *Bathybius,* but did raise the possibility, at least, that life began deep, not shallow; if not in abyssal ooze, then perhaps in abyssal fire.

A mile down in black, frigid waters off the Pribilof Islands of Alaska, a tiny warning bell chimes inside the sleek, grotesque head of a Baird's beaked whale. The whale ignores it and continues hunting. The Baird's beaked whale, *Berardius bairdii,* looks like a creature from a medieval bestiary. It has a smallish, dolphin-like head attached to the big body of a whale. Its long beak has a duck-like upward curve and a pair of short tusks near the tip. The whale grows to 40 feet and lives on the deep-sea fish and cephalopods of the North Pacific.

The bell chimes again. The carbon-dioxide buildup during the dive causes gas sensors in the whale's tissues to warn it to return to the surface. The whale chases down another squid. Two more squid flee ahead, and the whale, swallowing, tracks them using echolocation. It disregards the noisy drumming of several brotulid fish. Brotulid and rattail fish have drums in their swim bladders—tympanic structures that humans have dissected out, but have never heard. The Baird's beaked whale knows the sound. The whale ignores the familiar background murmur of the deep, concentrating on its own click-trains rebounding from the squid. The bell chimes again, more insistently. The whale spares the rest of the squid. They blur, then fade on its biosonar, and the whale starts up to the top.

Beneath the snow-capped pinnacle of Japan's sacred Mount Fuji lies 40-mile-long Suruga Bay. A few miles from shore, the seafloor plunges steeply, harboring in its depths an astonishing array of marine life fed by nutrients from the Kuroshio current and surrounding mountains. Only recently—with the development of ROVs (remotely operated vehicles) and advanced deep-water cameras—have many of the creatures been photographed in nature for the first time. At 450 feet (right), a six-foot-wide giant spider crab is captured by the National Geographic's SeaROVER in its partner MiniROVER's lights. Two feather stars (below)

Cyllometra manca, Crinoid

Macrocheira kaempferi, Giant Spider Crab

Scorpaena onaria, Scorpionfish

at about the same depth display fringed rays whose sticky tube feet trap minute organisms.

A unique shot made from *Nautile,* a French submersible equipped with National Geographic cameras, shows a chimaera (left) lured by tuna bait at 1,450 feet. The bizarre, shark-like creature trails two egg cases, which, once released, will feed the fish's embryos until they hatch months later. Also called ratfish for its long, pointed tail, the chimaera has only one set of gill openings and an upper jaw that is fused to its skull. It flaps its pectoral fins when swimming like a bird flying in slow motion.

The patterning of a bottom-dwelling scorpionfish, photographed at about 450 feet (above), helps it blend in with its surroundings, the better to ambush smaller fish. This perch-like creature, part of a family of 300 species, some venomous, has both head and fin spines; the latter can inflict deep wounds.

With a camera that makes close-ups by remote control, SeaROVER

Gymnothorax sp., Moray Eel; *Maurolicus muelleri,* Bristlemouth

caught a cutlassfish darting past a tube anemone at 750 feet (right). Here, where sunlight is too dim for plant photosynthesis, most animals ascend to feed at night. The cutlassfish and many other twilight zone denizens know it is time to rise when their brains sense darkness through translucent skin patches on their heads.

The ROVs' bright lights attracted a school of bristlemouths during a night dive at 450 feet; the fish, in turn, attracted a moray eel, which promptly ate them (above). Another of the leathery, snakelike morays—one of ten eel species in the bay—charged SeaROVER, but even with powerful fangs, it couldn't get a grip.

On the seafloor at many depths lives the secretive brittle star, a spiny five-armed creature of a class that also includes sea stars and sea urchins. Unlike their more rigid cousins, brittle stars have thin, sinuous arms that enable them to move rapidly. This mobility and their small size and deftness at hiding out in shadows and crevices make them highly successful inhabitants of the oceans' dimmer regions.

Six-inch serrated arms that can grasp and, like an elephant's trunk, carry food to its mouth characterize *O. tumida* (below). If lacking shelter, this brittle star can be found exposed on the ocean floor between about 800 and 1,800 feet, where light barely penetrates. *A. tenue* (opposite) has prickly arms as long as a foot and a half. They wrap around a host like this sea whip, forming loops. Tiny food particles adhere to sticky spines on the arms; tube feet then pass the particles along bucket-brigade-style to the brittle star's mouth.

Below the range of sunlight, about 3,300 feet, the

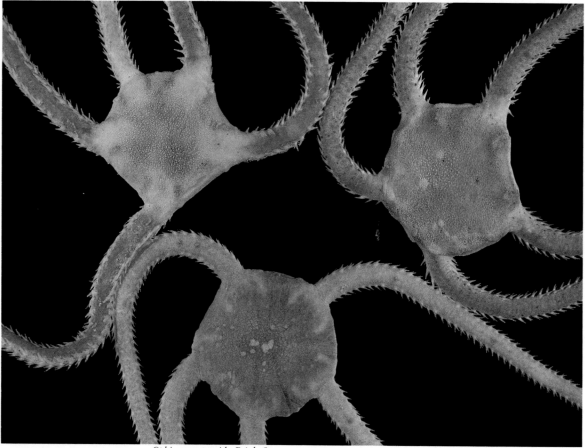

Ophiomyxa tumida, Brittle Star

Asteroschema tenue, Brittle Star

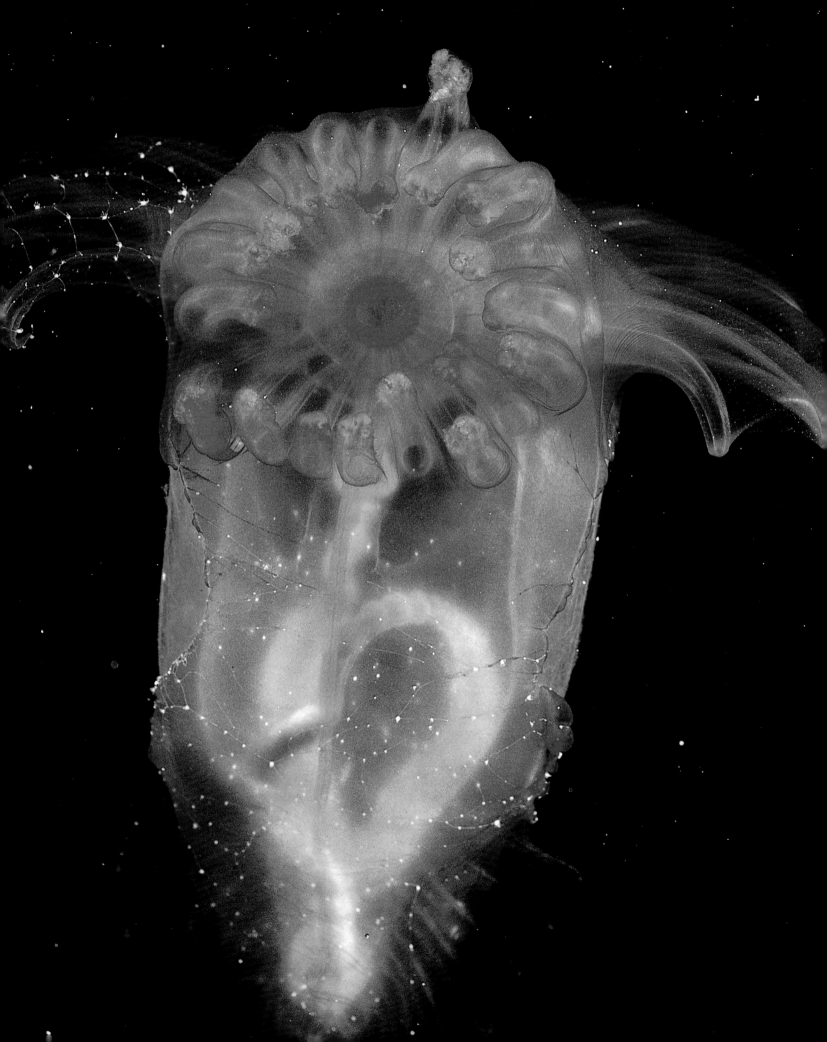

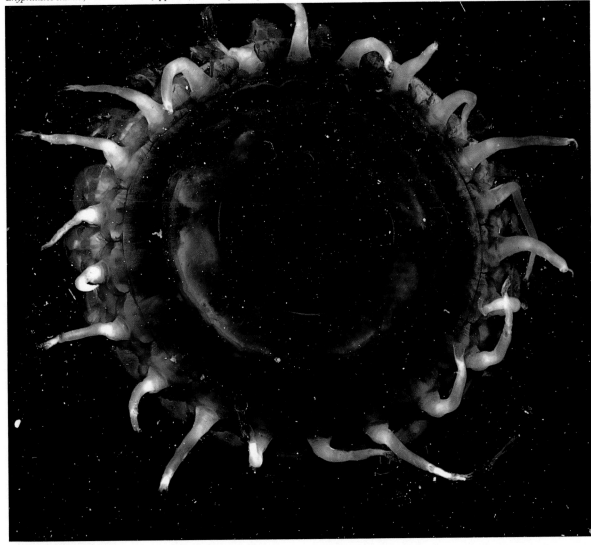

ocean becomes a black, cold, and sparse place. Earth's largest habitat—and more than 80 percent of the ocean realm—exists below the illuminated upper layer, yet no plants and few animals survive there. One creature that not only survives but flourishes is the sea cucumber (opposite), which can grow a foot or more in length.

The gelatinous, bottom-feeding animal traps food in its tentacles, then stuffs it bit by bit into its mouth.

Stinging tentacle cells on another large animal, the deep-sea jellyfish *Atolla* (above), allow it to snare, paralyze, and feed on sizable fish. Like other jellyfish, *Atolla* is more than 95 percent water; it drifts with the currents

and rhythmically pulsates to maintain position at the proper depth.

Pages 60–61: At about 2,600 feet, an 8-inch shrimp (whose tufted head indicates infection) feasts on one of its smaller kin.

Pages 60–61: Pasiphaeidae, Deep-sea Shrimp

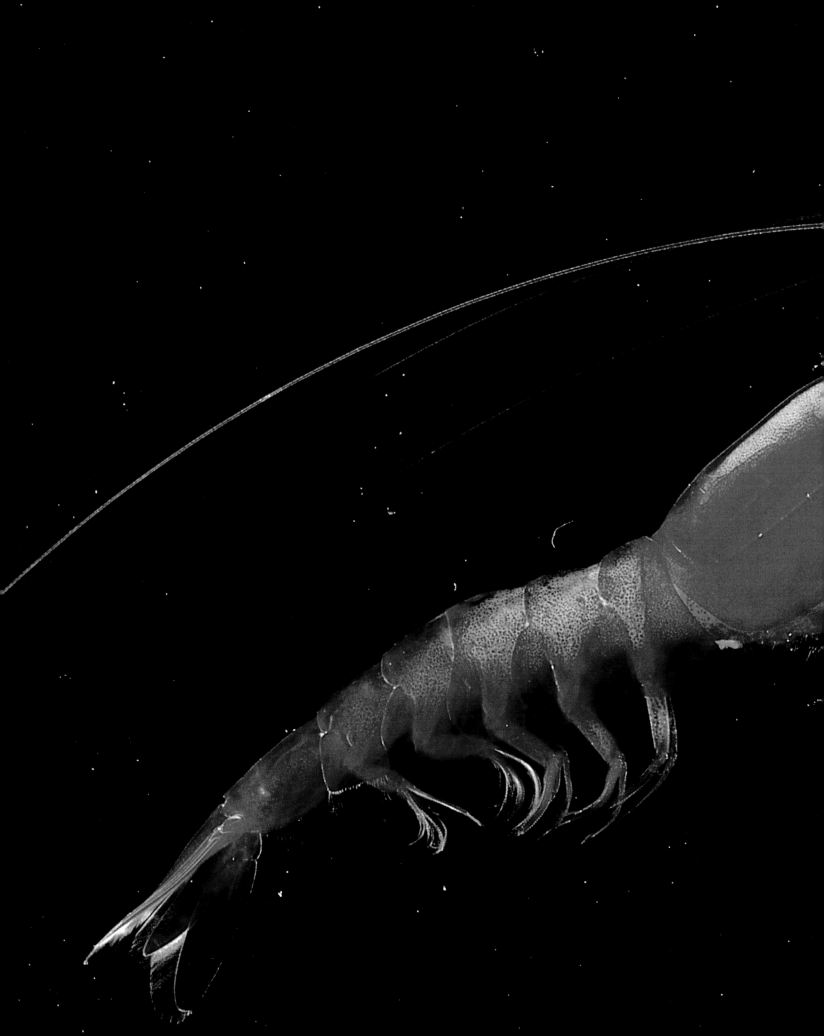

Saccopharynx sp., Gulper Eel

Siphonophore

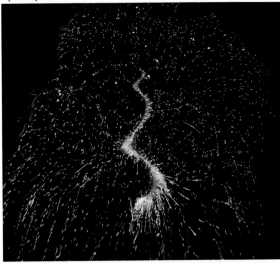

Alloposus mollis, Pelagic Octopus

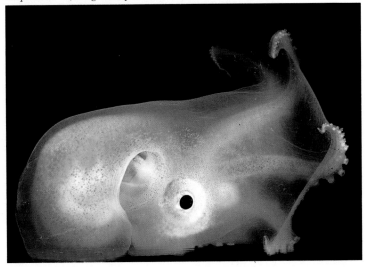

Nearly all mouth and tail, this gulper eel scavenges in the depths. Outsized jaws and well-developed teeth enable it to capture and eat a wide range of food, including deep-sea cod and other fish larger than itself. An elastic stomach conveniently expands according to the day's catch. In contrast, a delicate and complex colony of gelatinous creatures called a siphonophore (top), here lighted by a strobe, spreads itself like a spider web. Once paralyzed by stinging capsules, prey is passed to the central mouth, digested, then pumped to other parts of the colony. The beady-eyed octopus (above) captures prey with arms enclosed in a sheath.

OASES AT THE VENTS

Led by powerful floodlights, a research submarine, *NR-1,* cuts through the cold, foreign world of the deep. Unlike conventional submarines, the nuclear-driven Navy vessel stays submerged for weeks at a time exploring mid-ocean ridges and searching for a phenomenon new to science: hydrothermal vents on the seafloor.

Spewing mineral-rich water at temperatures above 650°F, such vents occur along many stretches of the oceans' 46,000-mile rift system. On a 1977 dive to the Pacific's Galápagos Rift, scientists aboard the research submersible *Alvin* were astounded to find collections of exotic animals around the vents. There, one and a half miles down on the desert-like ocean floor, giant clams, tube worms, a tube anemone (below), and other large, unusual species flourished. Since then, *Alvin* has carried scientists into numerous other rifts in both the Pacific and Atlantic; the vent communities they have found include some 25 previously unknown invertebrate families and subfamilies.

Tube Anemone

Riftia pachyptila, Tube Worms

Bacteria fueled by hydrogen sulfide from the ejected vent water (opposite) form the community's basic food source. So large is the concentration of food around the vents—as much as 500 times greater than elsewhere on the ocean floor—that some tube worms grow almost five feet long (above). These peculiar animals live symbiotically with dense bacteria colonies inside their bodies.

Pages 68–69: Foot-long clams live in mutually beneficial relationships with the bacteria in their gills. The clams absorb hydrogen sulfide through their feet; their blood carries it to the bacteria, which convert it into sustenance for the clams.

Pages 68–69: *Calyptogena magnifica*, Giant Clams

Polar Seas

A Baird's beaked whale, rising from its northern deeps, passes the luminous lure of an anglerfish. The light flickers, lonely, like a nomad campfire on a dark plain. The angler twitches its lure alluringly. The whale is not charmed. Hastening upward, it scatters the running lights of a school of lanternfish. The pattern of cold lights on each flank is repeated exactly in every other fish of the school. The whale passes, its turbulence dissipates, the lights of the school reform.

As pressure decreases, the whale's lungs expand and it gains buoyancy. The same hydrostatic laws that sped its descent now speed its return to the surface. The ocean blackness gives way to a dim gray, then the gray rises to a weak blue-green ambience. The Baird's beaked whale surfaces at 60° north on steep seas. It blows in the endless daylight of northern summer, its warm breath condensing explosively in cold arctic air. A wedge of sooty shearwaters divides and banks, skimming off over the steep slope of the swell. Gratefully, into its great lungs, the Baird's beaked whale sucks raw, wild air from one of the ends of the earth.

A world away, off the South Shetland Islands at 60° south, an Arnoux's beaked whale surfaces on steep southern seas. It blows explosively in the endless night of southern winter, its breath condensing in cold antarctic air. A wandering albatross veers away. The curtains of the aurora australis—the southern lights—dance overhead.

The Arnoux's beaked whale, *Berardius arnuxii,* is the Baird's beaked whale's southern cousin. It is not absolutely certain that the two whales are not the same species. Their populations are now geographically isolated, however, and if not separate species, they may be in the process of becoming so. The Arnoux's seems to be slightly smaller than the Baird's. It lives on the deep-sea fish and cephalopods of the Southern Ocean, the cold circumglobal moat that surrounds Antarctica. On recovering its oxygen debt, the Arnoux's beaked whale dives again. The cold lights of the aurora australis fade behind it. The cold lights of the bioluminescent deep begin to glow ahead.

Commonly known as cape pigeons, this petrel species breeds only in the Antarctic and subantarctic. Here a group floats near the South Orkney Islands, northeast of the Antarctic Peninsula. The gregarious birds often follow ships in huge, noisy flocks. Their chief food is krill, which they peck from the water in pigeonlike fashion.

The two whales inhabit mirror universes. The one universe lies under the North Star and the pale, shifting curtains of the aurora borealis. The other lies under the Southern Cross and the eerie veils of the aurora australis. In the one swim the northern elephant seal, the northern fur seal, the northern right whale. In the other swim the southern elephant seal, the southern fur seals, the southern right whale. In the one, the gull-like parasitic jaeger has filled the niche of hawk. In the other, another gull-like seabird, the south polar skua, has filled the same niche. In the one, the ultimate carnivore is the killer whale. In the other, the ultimate carnivore is the same animal. Both universes are hunted by the sperm whale. Both are grazed by blue, fin, humpback, and minke whales. Each of the latter big plankton-grazers looks identical to its analogue in the other universe, yet it is not—not quite. There is a tiny but momentous internal difference in programming. All the whales have the same migration pattern: poleward in summer, equatorward in winter, *but in contrary summers and to opposite Poles.*

If the higher animals seem to mirror their polar opposites, then so do the lower. Both Arctic and Antarctic seas show examples of gigantism among their simpler, bottom-dwelling creatures. Many polar sea spiders, sea lice, sponges, and foraminifera grow to huge size. Both faunas are characterized, too, by the phenomenon called polar emergence, in which species dwelling in the cold depths of the tropics and subtropics live close to the surface at the Poles.

Arctic and Antarctic are not perfect reflections of each other. Not all the animals in one have counterparts in the other. The Arctic is named for the constellation Arktos—the bear—the animal the Greeks saw in gazing north at that group of stars. By curious coincidence, the ice of northern seas does have a bear—the only seagoing bear on earth, *Ursus maritimus,* the polar bear. If "Arctic" is doubly apt, then the Antarctic, too, is well named, for it is both the antipodes of the Arctic and without a bear.

The Arctic, for its part, has no equivalent of the leopard seal. The leopard is the only phocid—or "true" seal—with warm-blooded prey, feeding regularly on penguins and occasionally on other seals. The absence in the Arctic of such a seal may have something to do with the capabilities of the seabirds there. The penguins of the north—the auks, auklets, puffins, murres, and guillemots—are wonderfully insulated by feathers, like their southern counterparts, and they dress in the same shades of white, black, and gray, and they mass in the same noisy colonies, but in addition they fly. For a bird-eating seal, that would be discouraging. The leopard seal's favorite strategy is to cruise the ice front before penguin rookeries, waiting for penguins entering or leaving the sea. This would never work with a tufted puffin, passing homeward like a bullet overhead. That the seabirds of the Arctic have retained the power of flight may be in turn explained by the presence there of terrestrial predators—the weasels, foxes, wolves, and bears that are entirely missing from Antarctica.

The Antarctic also lacks small, locally resident toothed whales like the beluga and the narwhal of the Arctic. It has never had an equivalent of a Steller's sea cow, a seaweed-eating northern dugong relative that once inhabited the Bering Sea. A Steller's sea cow grew to 25 feet and 7 tons. It was slow, curious, and tasted like veal—a fatal combination of traits. By 1768 it had been hunted to extinction.

The phocids are more diverse in Arctic than in Antarctic waters. Where the northern ice has ribbon, ringed, spotted, bearded, hooded, harbor, and harp seals, the southern ice has Ross, Weddell, crabeater, leopard, and southern elephant seals.

The Antarctic, however, has the numbers. Antarctic seas are among the planet's most fecund. There are fewer species of seal, but one of those species—the crabeater—is the most numerous seal on earth. The shrimplike crustacean on which the crabeater feeds, *Euphausia superba,* Antarctic krill, is one of the most numerous animals on earth.

There is a varied clangor under polar seas. In addition to the drumming of deep-sea brotulid fish, the bangs of sperm whales, the whistles and click-trains of killer whales, and the singing of humpbacks—normal background noise in any sea—there are, in northern waters, the knockings and gongs of diving and surfacing walruses, the extraterrestrial whistling of narwhals, and the burps, chirps, moos, mews, snorts, snores, screams, grunts, bellows, whinnies, whistles, trills, tooth-grindings, bell-chimes, rusty-hinge creaking, and outboard-motor imitations of belugas. The beluga, called the "sea canary" by the old whalers, has a vocabulary of several hundred sounds. It is the most voluble of cetaceans and hardly ever shuts up. At either end of the planet, too, can be heard the haunting, indecipherable pings and trillings of seals, the unearthly music of the planet's cold seas.

Geophysically the two Poles are less reflections of each other than negatives. Antarctica is a continent encircled by ocean. The Arctic is an ocean encircled by continents and large islands. The greatest height of Antarctica, the Vinson Massif at 16,860 feet, makes a match with the greatest depths of the Arctic Ocean, about 17,000 feet.

This contrariness makes for big climatic differences. The weather over the Arctic Ocean is influenced by the warming and cooling of the mainlands and large islands that rim it. The Arctic climate, as a result, is warmer and more seasonal than that of the Antarctic, and terrestrial life is richer. Antarctica, almost entirely covered by thick ice, never heats up.

Strong, cold winds blowing down from the high polar plateau cool the surrounding ocean. The Antarctic marine ecosystem is more isolated, colder, and more prolific than that of the Arctic.

The most striking of adaptations in polar sea animals is to the cold. In the chill of the Arctic and Antarctic, as in the chill of the abyss, the sperm whale is warmed by what Herman Melville called "The Blanket" in a *Moby Dick* chapter by that name. The sperm whale, he noted, may be covered by eight inches of blubber. "It is by reason of this cosy blanketing of his body, that the whale is enabled to keep himself comfortable in all weathers, in all seas, times, and tides. What would become of a Greenland Whale, say, in those shuddering, icy seas of the North, if unsupplied with his cosy surtout?"

The blue whale is warmed by four inches of blubber. The beluga keeps warm under the insulation of two inches of it—40 percent of its body weight. The Weddell seal is warmed by up to four inches of fat.

The walrus beats the cold with three inches of blubber and by vasoconstriction in its exterior. When the walrus dives, vascular constriction of the vessels in its fat and skin shunts blood inward. The skin, unheated by blood, quickly cools until its temperature is nearly that of the surrounding sea. The blubber, drained of blood, becomes an insulating blanket. When the walrus hauls out again, the blood is shunted the other way. (In air, for an animal as massive and well insulated as the walrus, the problem is not keeping warm but cooling off.) The walrus vasodilates, and blood once again suffuses its skin.

The walrus is not alone in this sort of thermoregulation. All mammals are capable of it. The white bellies of tropical dolphins glow rosily as they dissipate heat by vasodilation. The skin of a human swimmer is cold and pallid from vasoconstriction as he emerges from the water. The true seals are masters of the thermoregulatory art. But the phenomenon is especially striking in the walrus. On first hauling out, the walrus looks pale and bloodless. Then it vasodilates and flushes to a reddish brown.

The problem of maintaining warmth is most acute for young animals. Bulk conserves heat, and newborn sea mammals have least of it. Harp seal pups are born on sea ice in the heart of winter weighing no more than 20 pounds. The famous whiteness of the pup's pelt is one of its adaptations to cold. The hairs, on close analysis, prove to be translucent. The sun shines through, and its heat is trapped close to the skin—a small, local greenhouse effect. Like the coat of the polar bear, the pup's coat absorbs much of the ultraviolet spectrum in sunlight. Photographed aerially with special ultraviolet-sensitive gear, harp seal pups appear as dark shapes against the whiteness of the whelping ice.

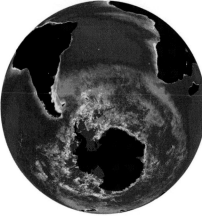

Spring brings long days of sunlight to the Poles, causing an explosive bloom of phytoplankton. Satellite images of the Arctic (upper) and Antarctic (lower) show the heaviest concentrations in deep red, shading down through the spectrum to violet. Steel gray areas represent an absence of data.

Pages 76–77: A research schooner (upper left) skirts icebergs and pack ice off Antarctica. The world's largest icebergs are those that break off from Antarctic ice shelves.

Another adaptation to cold in harp seals, as in other sea mammals, is the richness of the mother's milk. Harp seal milk is about 45 percent fat, 10 percent protein. Within two weeks of its birth, the pup weighs 60 to 80 pounds. The transfer of mass from mother to pup is speedy and efficient. A female harp seal weighs about 400 pounds. In less than a month she has given nearly a quarter of herself. Nursing sessions on the harp seal whelping ice are a race against the day when the pup must enter the sea.

If true, or "earless," seals like the harp seal have one set of strategies against the cold, the "eared" seals, or otariids—the fur seals and sea lions—have another. In both families of seals, the hair grows in bundles. Each bundle has a single flattened guard hair protecting a tuft of underhairs that trap air and serve as insulation. In true seals, the phocids, there are from two to five underhairs under each guard hair. In fur seals, hairiest of otariids, there are as many as twenty underhairs.

One disadvantage for fur seals is that thermoregulation becomes more difficult. In fur seals, vasodilation works only in the flippers. Everywhere else, fur traps the heat the animal needs to dissipate. To cool off on land, a fur seal must wave its hind flippers in the air. A harp seal simply relaxes the arterioles in its fat and skin, which fill with blood and release heat.

Superior thermoregulation is one reason, perhaps, that the phocids are more successful. There are 15 times as many individual phocids as there are otariids. They have adapted to more varied habitats, especially in the planet's colder seas.

Of the two Poles, the Antarctic is the more otherworldly. While the sources of the Arctic Ocean, both for its waters and its species, lie largely in the Atlantic, with a smaller contribution flowing north from the Pacific through the Bering Strait, the Antarctic makes its own world. It is a world circumscribed and defined by currents. Nearest the Antarctic mainland, easterly winds set one surface current—the east wind drift—flowing west. Farther out, westerlies set another current—the west wind drift—flowing east. This outer, eastward current, also called the Antarctic Circumpolar Current, is the largest on earth and the only current to circle the earth unobstructed by landmasses. It is a thermal barrier to creatures from warmer latitudes.

At the edge of the west wind drift is the Antarctic Convergence, the meandering border that divides Antarctic from subantarctic. The west wind drift ecosystem lies north of the limit of winter pack ice. Its primary producers are phytoplankton, which are eaten by planktonic crustaceans, which are eaten by lanternfish, which are eaten by squid, which are eaten by sperm whales and beaked whales.

The east wind drift ecosystem, the more productive, is based both on phytoplankton in the water and on algae growing in the ice. It is the zone of another current circumscribing and defining Antarctica—the living river of krill. Antarctic krill is the largest species of euphausiid shrimp. It is long lived and in many ways behaves more like a fish than a crustacean, schooling like a sardine and spawning several times a season. Krill is the keystone species in the Antarctic food web; staff of life for penguin and petrel, crabeater and leopard seal, minke, fin, and blue whale.

When Antarctica was of a piece with Africa, Australia, South America, and India—just another territory in the old southern supercontinent of Gondwana—and for a time after drifting away with Australia, it was home to ferns, beech trees, deciduous conifers, and big reptiles like *Lystrosaurus*. Its coastal waters supported sharks, rays, and catfish. With its steady drift south, and with the formation of the Antarctic Convergence, species began dropping out. The chilling of Antarctica was a complex and circular phenomenon. One theory holds that, as the continent migrated deeper into southern winter, winds blowing down from the growing polar ice cap cooled the surrounding currents, which in turn blocked any warming by currents from the north. For life on land, it was a spiral down the thermometer into a frigid perdition. The continent froze over. All terrestrial mammals, reptiles, and amphibians died off. Most land birds and temperate fish perished.

Seabirds prospered. Life in the interior was now extinct or nearly so. Almost all food was in the ocean, and the seabirds were already adapted to hunting for that. They were accustomed to cold weather, as well. One order, the "tube noses," made the most of their circumstances. They diversified—four species of albatross, six of fulmar, various shearwaters, prions, storm petrels, diving petrels—and proliferated to become the most numerous of Antarctic birds. Penguins, too, made the best of a cold thing. The penguins, like the petrels, were preadapted by their insulation. As Antarctic winter deepened, the penguin's layer of fat, its woolly undercoat of down, and its outer parka of overlapping feathers deepened too.

Some lower animals in Antarctic seas were able to cheat the cold by their simplicity. Sea spiders and sea lice have a single body cavity filled with fluid as salty as or saltier than the sea, and the salts depress the normal freezing point. Fish biology is more complicated than that. Ice easily penetrates fish gills, mouth, and skin, and because fish are cold-blooded, with body temperatures close to that of the environment, the ice can quickly propagate inward and freeze them. Any ice crystal, internalized, can become the seed of a fish's destruction.

One group of fish, the suborder Notothenioidei, thrived. These perch-like fish radiated to fill the niches of the temperate fish frozen out, and today about two-thirds of Antarctic species are notothenioids. One secret of

79

their success is chemical. The notothenioids produce antifreeze. In warmer seas, marine fish follow the sea louse strategy, depressing the freezing point by means of salts in their body fluids. In the notothenioids of the Antarctic, salts account for only 40 to 50 percent of freezing-point depression. The rest comes from eight different antifreeze molecules—glycopeptides. Each ice crystal that enters any part of the fish must be coated with antifreeze.

The blood of some Antarctic fish is thin and pale—oxygen-carrying red blood cells are not essential, because oxygen is highly soluble in cold seawater—and this thin, less viscous blood helps the fish to lower their metabolism, saving energy in an environment where energy conservation is vital. Modifications to achieve neutral buoyancy—a light skeleton and fat for flotation—are another energy-saving strategy.

The seals, themselves preadapted to cold seas, were fruitful and multiplied, diversifying across a number of niches. The crabeater seal lives offshore on the Antarctic pack ice. It has strange, lobed teeth that mesh to work as a plankton strainer. Feeding directly on krill, at the bottom of the food chain where biomass is greatest, crabeaters have become the most numerous of seals. There are likely more of them now than ever. Their numbers have grown to help fill the void created when Antarctic whalers devastated the populations of humpback, fin, and blue whales, the biggest krill-eaters of all.

The Ross seal is a second inhabitant of the pack ice. It is a smaller seal but hunts larger prey—midwater squid, fish, and krill. It has enormous eyes for hunting the dim waters beneath the floes. The eye of the Ross seal, like that of all seals, is a wonderful bit of optical engineering adapted to a variety of polar exigencies. It has a nearly round, light-gathering lens and a reflective, light-amplifying tapetum—a membranous layer behind the retina. It has powerful dilator muscles that open the pupils wide, and a strong corneal astigmatism to correct for underwater distortion. When the Ross seal returns in daylight to the glare of summer ice, the huge pupils narrow to slits, stopping down the aperture. The slits lie parallel to the astigmatic axis, which minimizes the astigmatism.

The leopard seal is a third seal of offshore ice. Its niche is somewhat broader. Leopard seals are opportunistic eaters. Examination of their stomachs reveals a little of everything: penguins, fish, squid, clams, seals, sea lions, and once—somehow—a duckbill platypus. But most of their diet is krill. For leopard seals, as for most big Antarctic predators, those shrimplike crustaceans are the staple of life. The leopard's omnivorous habit is reflected in its teeth. Its canines and incisors are wolflike, suited to taking large prey. Its cheek teeth, each with three cusps that interlock with the cusps of the opposing teeth, are suited to straining krill.

The Weddell seal is the southernmost and most Antarctic of seals. It dwells on, and mostly under, the shore ice of the continent. The adaptations allowing it to inhabit permanent ice are its superb insulation and a wide gape. The gape permits its protruding top teeth to carve away at the ice forming in its breathing holes, keeping them clear. This is an endless chore, and the teeth of Weddell seals are much worn in old age. Weddells were once thought to be the deepest divers among seals. Their need to return to the breathing hole made it easy to attach depth recorders to them. After the seals had done their involuntary bit for science, it was easy to remove the recorders again. Elephant seals have since been found to surpass them, but Weddells still hold second place, with recorded dives as deep as 1,900 feet and as long as an hour. Oxygen for that hour is stored not in the lungs—a seal's lungs collapse when it dives, preventing the bends—but in hemoglobin in the blood and in myoglobin. The latter is a dark substance so abundant in seal muscle that it turns the muscle nearly black.

The real harshness of polar life, Arctic or Antarctic, is less in the cold than in the dark. No heat-exchange system, no thickness of blubber, can feed an animal in the long, unproductive winter of lightlessness. A common strategy against the dark, and the cold as well, is simply to flee it.

As days grow short in Antarctica, Adélie, chinstrap, and gentoo penguins head north for the pack ice and beyond. The skuas and all the great family of petrels disperse northward. Arctic terns start the long flight for their nesting grounds in Alaska, Greenland, and elsewhere in the north. The birds leave the continent to emperor penguins, the only ones of their kind to winter there.

Any seals dallying inshore scatter offshore into the pack ice. They leave the continent to the Weddell seals. The Weddells themselves flee the cold, but their flight is downward just three to nine feet. They spend much of the winter in the dark waters under the shore ice. Winter temperatures in Antarctica can reach −130°F and combine with gales blowing down from the polar plateau to make a horrendous windchill factor. The icy water is cozy, compared to that.

By the middle of March, with night skies dark enough for the southern lights to show, and with slush ice forming offshore, a blue whale dives shallowly for its last meal of the season. The plankton bloom is fading, the water growing clear again. One hundred feet down, in the dim blue light, the whale appears as mottled blue-gray on top, pale underneath. Its 90 feet of underside are mustard-colored from diatoms accumulated during its stay in cold Antarctic waters. On the surface, it had seemed a slender whale. Now, undersea, on opening its great mouth, expanding its ventral pleats, and filling with water and krill, it swells into a blimp.

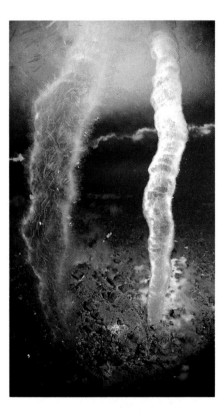

Antarctic sea icicles (above) hang down seven to ten feet from the frozen surface of McMurdo Sound. On the floor, sea stars and other creatures of the Antarctic continental shelf form part of a thriving benthic community. A diver (opposite) sinks into this blue otherworld beneath the Antarctic ice.

Pages 82–83: A frozen stash of trout, an Inuit dietary staple, marks the end of a fishing trip in Canada's Northwest Territories.

The Swedish naturalist Linnaeus, in a comic mood, named this whale *Balaenoptera musculus*—little mouse fin-whale. Blue whale and mouse, in fact, occupy opposite ends of the mammal spectrum. The blue whale is the largest creature that has ever lived. Before large-scale commercial whaling, there were about 200,000 of them in Antarctic waters; today there may be as few as 1,500. This survivor closes its mouth and expresses tons of water through the filter of its baleen. It grows slender again. With a tongue larger than a taxicab, it cleans 40 pounds of krill from the inside of the baleen—slim pickings, a last snack for the road. It surfaces, blows, and heads north for its wintering grounds off Ecuador.

As days grow short in the Arctic, there is a whispery thunder of wings as geese, pintails, eiders, oldsquaws, sandpipers, phalaropes, semipalmated plovers, great black-backed gulls, Iceland gulls, ring-billed gulls, kittiwakes, parasitic and pomarine jaegers, and common murres all lift off and head south. Arctic terns, in search of their endless high-latitude summer, start south for Antarctica. Their 22,000-mile annual migration is among the longest known. They are the ultimate symbol of polar life. There are a few other bipolar species, but Arctic terns are the only bipolar individuals—the only creatures that travel seasonally from one polar region to the other.

Belugas swim south for the estuaries in warmer waters where they bear their young. Walruses migrate down through the Bering Strait from their rich shellfish beds in the Chukchi Sea. Harp seals leave their summer feeding grounds above the Arctic Circle, swimming southward ahead of the advancing autumn sea ice.

Beneath the calm October surface of Icy Strait, Alaska, a humpback whale, *Megaptera novaeangliae,* swims a spiral upward, emitting dollops of air. The whale—a young bull—is spinning a "bubble net," concentrating copepods and small fish for a last meal of the season. The surface of the strait is glassy, for now, but a pair of migrating terns see drama coming and veer that way. Bubble clusters boil to the top, one after another, inscribing a circle on the water. Then the calm surface of the circle dimples with scatterings of small baitfish, then it erupts in the dark head of the whale. The humpback rises a third out of water, mouth agape, baleen showing. He falls back, mouth closing, to disappear in a mountainous white fountain of foam. He swallows, blows, turns southwest for the warm Hawaiian waters where North Pacific humpbacks breed and calve. Three thousand miles of open ocean lie between the whale and his destination. As he passes the sea-stunted spruces of Cape Spencer, he begins humming, at a hundred decibels, the stentorian bass notes and eerie trumpetings of the humpback's mighty song.

S trictly speaking, the term "Arctic" refers to the ocean and lands within the Arctic Circle, north of latitude $66\frac{1}{2}°$N. Ecologically, however, this makes little sense. Biologists place the Arctic land boundary instead at the tree line, north of which the average temperature of the warmest month is below 50°F. Oceanographers use temperature and salinity of the seas as their measure, and the marine *sub*arctic boundary matches at sea the tree line on land.

Around the permanently frozen polar ice cap, ice breaks up in summer. Land glaciers calve deep-keeled icebergs like the ones below, north of Greenland. Birds and marine mammals stream north in one of earth's great migrations, to feed and breed in Arctic seas.

Among species that spend their lives there are two small, gregarious whales closely related to one another—the talkative beluga (right), whose repertoire of sounds has been compared to an orchestra tuning up; and the narwhal (pages 88–89) with its long, spiral tusk.

Delphinapterus leucas, Beluga, 16 feet long

The tusk is the male narwhal's only functional tooth. In former times, Vikings and whalers traded narwhal tusk as the mythical unicorn's horn, prized for its magical powers. In particular, it was thought to neutralize poison. Narwhal tusk remains a collector's item, but the animals are now protected by environmental laws.

Tusks are also the trademark of walruses, such as the two above swimming beneath pack ice off Alaska. Both sexes grow them. In males—as perhaps also in male narwhals—large tusks help establish social dominance and may be employed as weapons.

Walruses also use their tusks to haul their bulky bodies out of the sea onto ice, inspiring their scientific name, which means "tooth-walking sea horse." At hauling-out places such as Alaska's Round Island (opposite), males bask companionably in the sun. Bullies shove and jab their way to favored spots in the heaving mass.

Walruses have many pinniped relatives—seals, sea lions, and fur seals—in both Arctic and Antarctic,

Above and Opposite: *Odobenus rosmarus divergens,* Pacific Walrus, 9.5 feet long

Ursus maritimus, Polar Bear, 9.5 feet long

but they themselves are found only in northern regions. So large are walruses, sometimes weighing as much as two tons, that they fear few predators other than humans.

Another Arctic native, the polar bear (left), may charge a walrus herd and pick off a laggard calf scrambling for water. The bear's chief food, however, is seal.

The most maritime of bears, polar bears spend much of their lives on the Arctic Ocean, lumbering across ice floes or swimming between them. Their dense fur is almost waterproof; together with a thick layer of blubber, it insulates them against the harsh Arctic climate. Even the bottoms of their paws are furred for warmth.

The mother at left may be leading her cubs on a hunting expedition. Their white coats will act as camouflage while the bears stalk prey or lie quietly in wait beside a seal's breathing hole.

The cubs have grown hugely from their birth weight of about one pound. They will stay with their mother for perhaps two years, learning the tricks of surviving in the Arctic. Then they will wander as solitary nomads until they grow old enough to mate. Males may eventually weigh more than half a ton, females less than half as much.

Ultimate symbol of the Arctic, *Ursus maritimus*— sea bear—bridges in name and behavior the land and sea realms of the far north.

In the sunlit upper layer of the seas float algae known as phytoplankton that sustain life in the marine ecosystem. These minute plants are the base of the ocean food web, being primary producers of organic material through photosynthesis. They are particularly abundant in waters of the Arctic and Antarctic (left).

At both Poles, massive spring phytoplankton blooms have their source in the cold, nutrient-rich waters and long days of sunlight. In Antarctica, scientists have discovered this is only half the story.

In winter, sea ice about 6 feet thick stretches seaward up to 1,000 miles around Antarctica. Much of this ephemeral extension

of Antarctica is a pasture of algae. In spring, algae living in and under the ice multiply. Krill—the chief zooplankton species of the Antarctic—are among the small animals that feed in all seasons on algae bearding the underside of the ice (below). These animals in turn become food for fish or squid.

After the spring thaw,

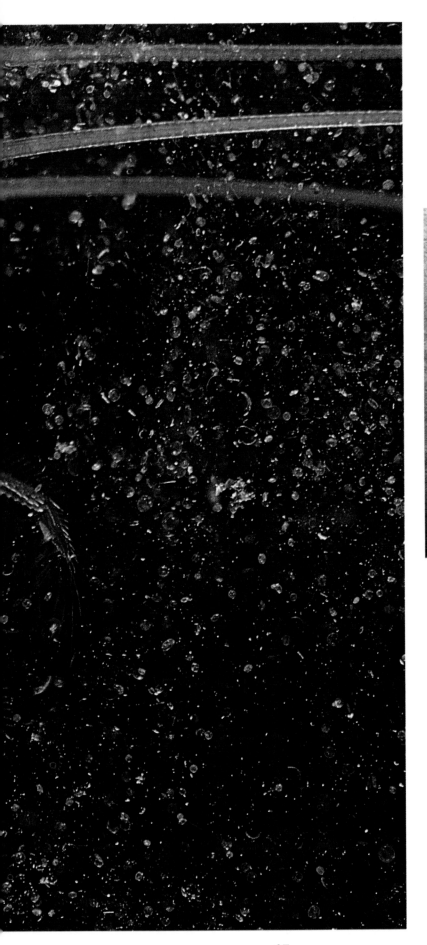

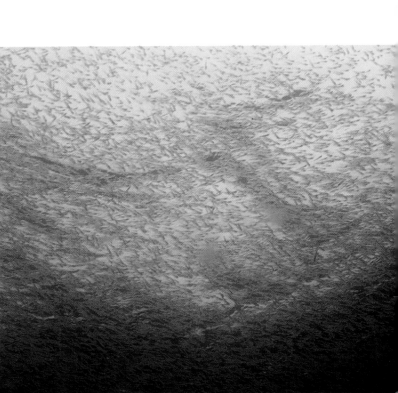

brine channels honey-
comb the ice, and these
are colonized by algae,
krill, amphipods, and
small fish. Even as the
spring sun powers the
phytoplankton bloom on
the ocean surface, it is
melting the ice and releasing the resident algae into
the sea. This profusion of
microscopic plants feeds
the multitudes of krill

(above), key animals in the
Antarctic food web.
    An Antarctic krill (left)
captures algal plants such
as diatoms in a sieve
formed by its hairy front
legs, filtering the plants
from seawater. Most Ant-
arctic animals feed on the
crimson tides of shrimp-
like krill or on other animals that eat them.
    Among other creatures

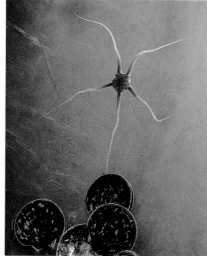

Ophionotus victoriae, Brittle Star,
7 inches in diameter

that graze on ice algae are bottom-dwellers such as brittle stars and sea stars. A long-legged brittle star (left) searches for algae growing on an underwater ice cliff. Sea stars also commonly eat algae during the phytoplankton bloom.

Supreme opportunists, sea stars scavenge for food on the ocean floor. Their chief targets are invertebrates, alive or dead. Above, several sea stars devour a dead mollusc in a slow-motion feeding frenzy.

Most sea stars eat by turning their stomachs inside out onto their food. They partly digest their prey outside their bodies, then finish it off inside.

Above this benthic community swim Antarctic fish, squid, and jellyfish. Amphipods often dot jellyfish (opposite), whose tentacles may trail four feet. Some amphipods on jellyfish are parasitic, but recent research shows that these are not. Their symbiosis may offer them protection from fish.

A major food source in the Antarctic, nourished by the abundant krill and other small prey, consists of

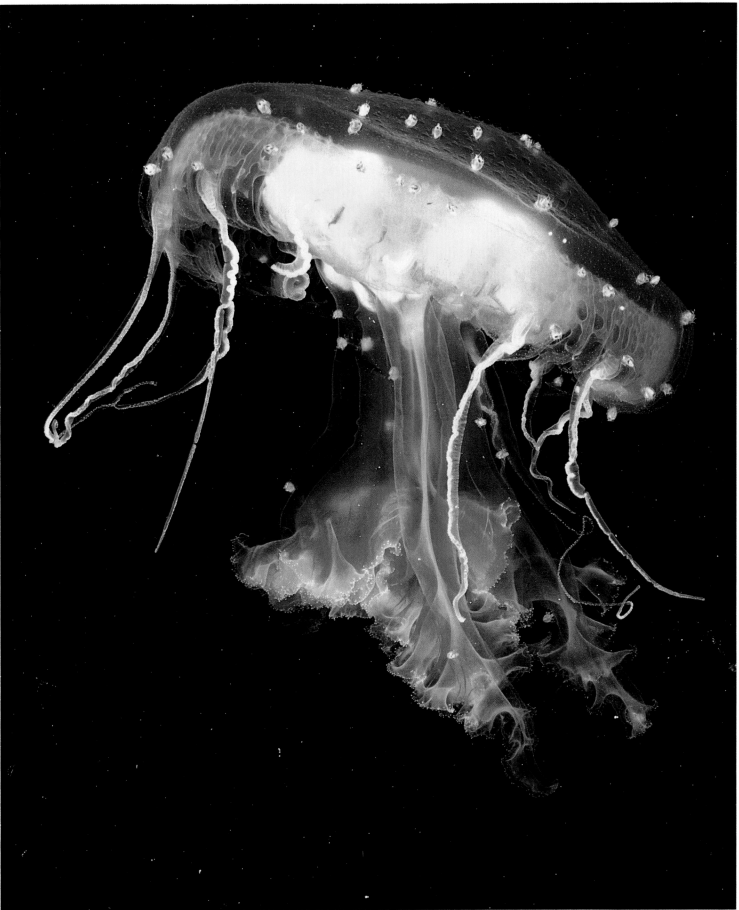

Trematomus bernachii, 7 inches long

squid and fish such as *Trematomus* (above). They are the primary food of the Bryde's whale (right), as well as of sperm whales, seals, albatrosses, and emperor penguins. Exploiting the riches of the food web, the killer whale eats other marine mammals and fish.

Most Antarctic baleen whales subsist mainly on krill, as do most petrels and other seabirds, penguins, and crabeater seals. They depend on the luxuriant bloom of phytoplankton in spring and summer to ensure a reliable supply of food for krill.

In recent years an environmental alarm has sounded. Tests indicate that depletion of the ozone layer over the Antarctic may be reducing photosynthesis in phytoplankton. A decrease in phytoplankton could seriously affect the food web.

In addition, krill are being touted as a protein source of great potential. Several countries already harvest hundreds of thousands of tons of krill each year. Biologists worry that an increased harvest might endanger the entire web of life in the Antarctic.

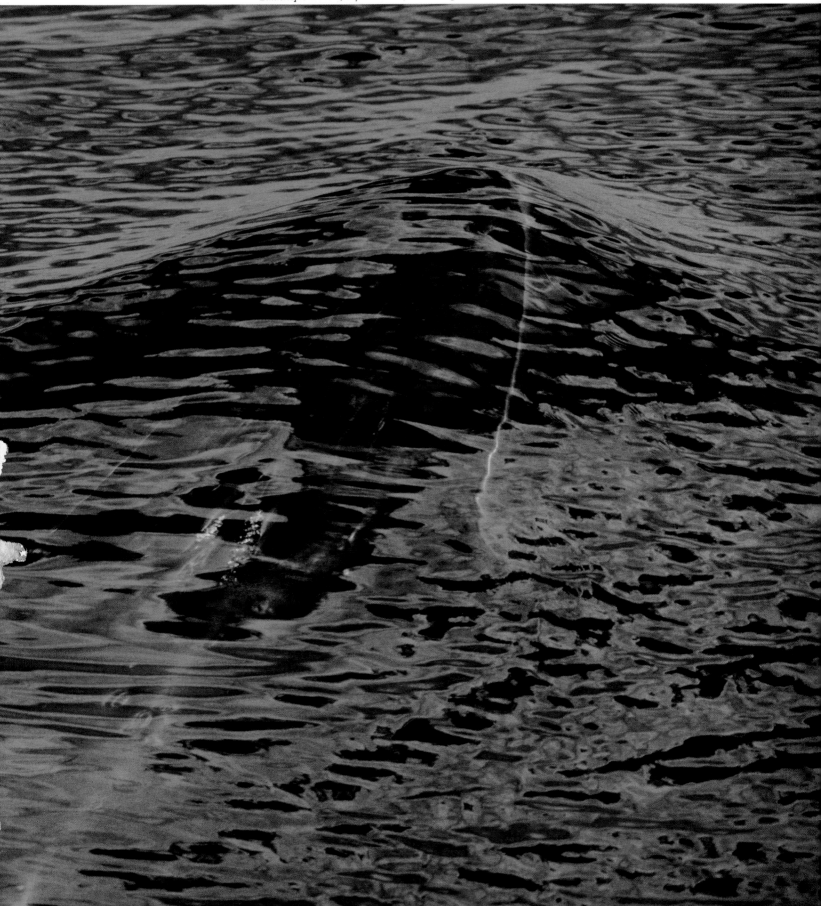

Balaenoptera edeni, Bryde's Whale, 45 feet long

Thoroughly at home on the Antarctic continent, Weddell seals (below and opposite) spend most of their lives on or under ice along the shore. With winter, other seals disperse onto the pack ice offshore, escaping the worst of the continental weather. Weddells find the relative warmth of inshore waters enough protection.

Weddell seals and other true, or earless, seals swim with strong lateral movements of their bodies, using their rear flippers for propulsion. The name "true" comes from the fact that these seals were the first to be described by European scientists. Their family name is Phocidae. "Earless" is something of a misnomer. The seals can

hear perfectly well through earholes in the sides of their heads. They simply do not have external earflaps, as do the eared seals, or Otariidae—sea lions and fur seals.

Other differences affect locomotion. Earless seals cannot bend their rear flippers forward to help in walking on shore, so they hunch along in caterpillar

Above and Opposite: *Leptonychotes weddelli*, Weddell Seal, 8.5 feet long

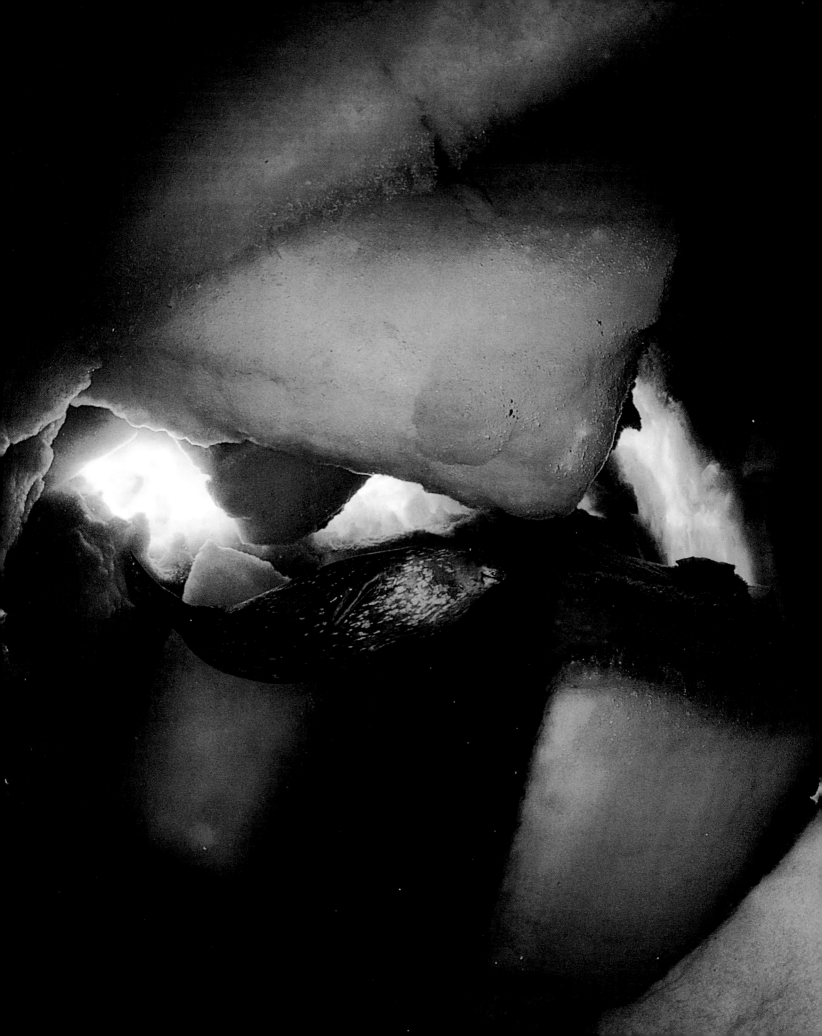

fashion. Eared seals can bend their rear flippers forward and raise their bodies on large, splayed front flippers, enabling them to gallop on all four limbs. They swim with their front flippers, using the rear ones as a rudder.

Most Arctic and Antarctic seals belong to the Phocidae family. Some phocids are bipolar, such as the elephant seals lolling peacefully above, or roaring in male challenge at right. These massive southern mammals are the largest of all seal species, with males sometimes attaining a weight of four tons. Females weigh a modest ton.

The male's "trunk" is a balloonlike proboscis that he inflates to ward off rivals. When he roars, the trunk forms an echo chamber, projecting the sound for half a mile. Battles erupt in the breeding season, as dominant bulls, known as beachmasters, defend their place in the hierarchy and their access to a harem of cows.

These powerful seals dive deep to feed on fish and squid. They are nearly matched in diving ability by the Weddells. All seals

Above and Right: *Mirounga leonina,* Southern Elephant Seal, 8 feet long (female) to 15 feet long (male)

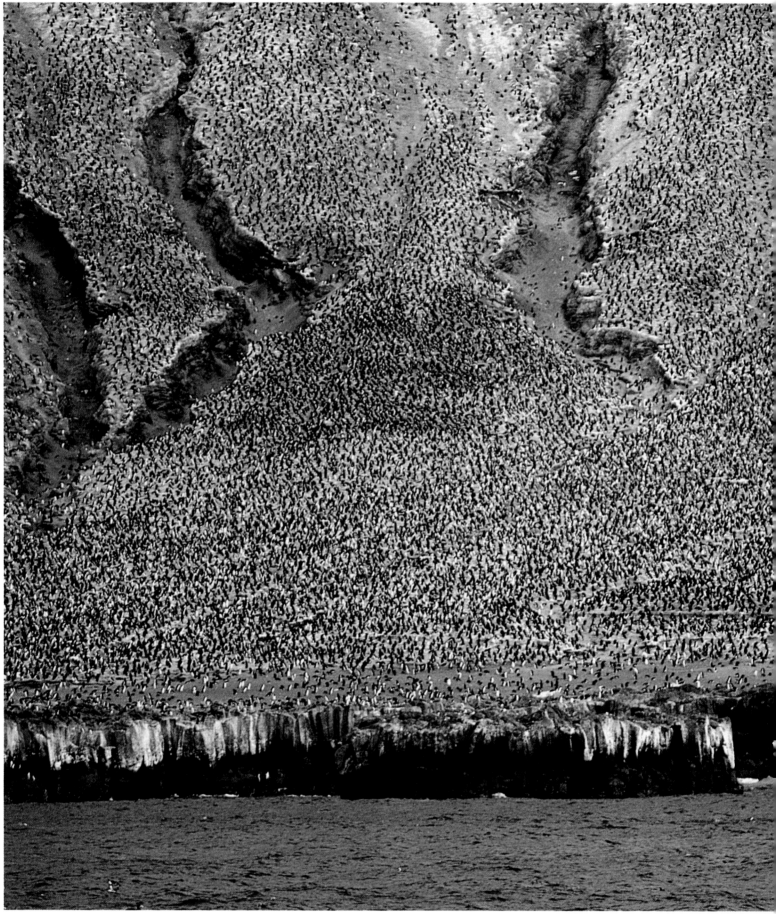

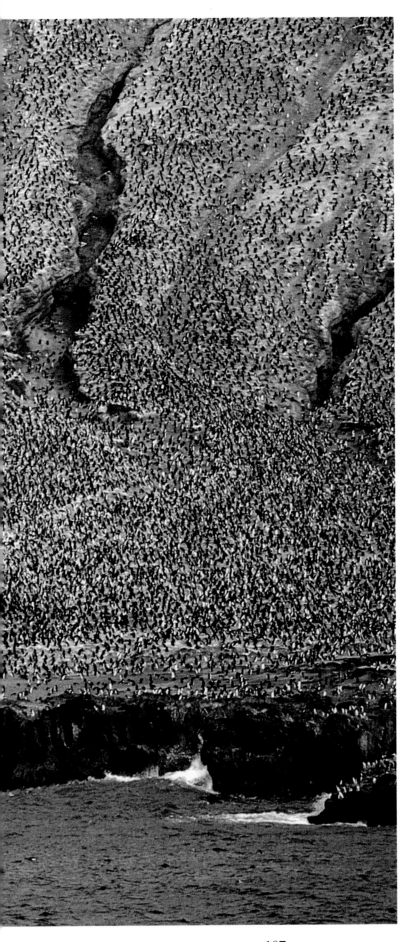

Pygoscelis antarctica, Chinstrap Penguin

are carnivorous, but they feed at different depths on fish, krill, or squid, and sometimes on penguins.

Seven species of penguin play their part in the food web of the Antarctic. These flightless birds are found only in the Southern Hemisphere, and have adapted in unique fashion to conditions at the South Pole. Dense plumage, with overlapping feathers and an undercoat of down, acts as insulation and waterproofing. A thick layer of fat helps protect penguins in and out of water and creates an energy reserve.

Highly gregarious, penguins eat, sleep, and breed in flocks. Their breeding colonies, known as rookeries, may contain hundreds of thousands or even millions of birds.

Crowded into a rookery in the South Sandwich Islands (left) are members of two different genera. The chinstrap is pygoscelid, or brush-tailed, with stiff tail feathers that sweep the ground. The macaroni is eudyptid, or crested, with striking orange-yellow head plumes.

Penguins have no land predators except seabirds

Catharacta sp., Skua, 22 inches tall

such as skuas (above, right). Scavengers of the penguin rookeries, skuas may carry off unguarded eggs and eat chicks or even disabled adults.

Sea predators such as killer whales and sea lions present more of a threat, but the most dangerous for Adélie penguins are leopard seals (opposite and above). These sinuous

creatures with curiously snakelike heads keep watch from the water below a rookery. Hungry but nervous Adélies often line up on shore—eager to go hunting for food but waiting for another bird to test the waters.

When it spots its quarry, the leopard seal lunges forward, its mouth gaping wide to seize the penguin.

The seal may shake the bird hard or dash it against the water to loosen and strip off the skin before eating the penguin. Leopard seals also prey on pups of other seal species, but a great part of their diet is a food they share with the penguins and most other Antarctic animals: krill.

Two penguin species larger than the others, the

Above and Opposite: *Hydrurga leptonyx,* Leopard Seal, 11 feet long; *Pygoscelis adeliae,* Adélie Penguin, 27.5 inches tall

emperors (opposite)—largest of all—and kings, dive down to hunt for fish and squid as well. Awkward on land, penguins are in their element in the sea. With flipperlike wings they "fly" underwater and swim strongly, porpoise-fashion, at the surface. Unlike other birds, their bones are solid, which improves their diving ability. During the breeding season, penguins hunt for food to bring back to their chicks.

In the austral spring, the gentoo (above, left) and most other penguins lay two eggs in ground nests made of pebbles and moss, and incubate them for about 35 days.

Emperors follow a different strategy. A female lays a single egg in winter, then leaves, and the male balances it on his feet, keeping it warm under a special fold of skin. Huddling with other males, he incubates it for 65 days without eating. His weight loss from courtship until the female returns at hatching time is about 40 percent. No other penguin is big enough to survive such an ordeal. Chicks grow fast on a diet of regurgitated food (above).

Pages 112–113: Most emperor rookeries are located on sea ice around Antarctica. As the ice breaks up in spring, the fledgling chicks float out to sea when the food supply is at its greatest, giving them the best chance of survival in their challenging world.

Open Ocean

pod of Dall's porpoises joins the migrating humpback. For a minute they ride the pressure wave preceding the whale, jockeying for position directly ahead of the knobby, barnacled, good-natured face. In that sweet spot, a porpoise catches a free ride, borrowing the whale's energy, gliding along without any effort of its own.

The Dall's porpoise is stocky and muscular, built as powerfully as a killer whale and painted in a similar striking pattern of black and white, but reaching just a fiftieth the weight. It spends more time dallying in the bow waves of boats and whales than do many of its cousins. When the young bull humpback slows to sing again, the porpoises lose interest and peel off to either side.

In Alaska's blue-green coastal waters, opaque with plankton, the dark surfaces of cetaceans quickly go invisible. Only pale markings catch the light. The last the porpoises see of the humpback are its white flippers. At 15 feet, adult humpback flippers are the longest in any species of whale. They inspired the humpback's genus name, *Megaptera*—Big Wing.

Through its summer months on the Alaska feeding ground, the humpback has not sung, yet now all the themes and phrases of its song come back to it. Humpback compositions are complex, rigidly ordered, sometimes 30 minutes long. The humpback brain is large, its memory for music elephantine. The song has survived through the summer's silence. Ten miles away, another migrating humpback clears its great pipes and begins to sing. The notes reach the young bull clearly, and they are familiar. Both whales are singing the same song.

Humpbacks are not great original composers. Themes invented de novo make for aberrant songs and may be just mistakes by young male humpbacks learning to sing or else the creations of older males seeking extra attention. All humpbacks of a given wintering ground sing the same song. The song is modified continually throughout winter, the singing season, but the changes come in reworking past themes, not in bold leaps of creativity. There are no Mozarts among humpbacks.

Revealing its identity, a humpback whale off southeast Alaska flings its flukes high in the air. Humpback flukes measure 10 to 15 feet across—about a third the whale's length. Broad and flat, the flukes' trailing edges are serrated; color markings, as individual as fingerprints, make it possible to track humpbacks whale by whale.

The young of Pacific spotted dolphins stay by their mothers for at least a year, often traveling in schools of hundreds. These in a group near Hawaii—one of three eastern Pacific populations—have fewer spots than those along North America's west coast. In all populations, spots, believed to aid in camouflage, increase with age.

The whale departs the continental shelf. The ocean deepens underneath it, and the water clarifies ahead. Sediment from rivers settles out; nutrients from continental upwelling decline; concentrations of plankton thin; and soon the water is a pellucid blue. The light grows cathedral-like. From the vaulted blue ceiling of the swells, long shafts of sunlight shine down. They come filtered as if through stained glass, but they never reach the nave. The drumming of bottom fish, the clicks and whistles of coastal dolphins, the whine of outboards, the plaint of buoys, the mumble of surf on headlands, all fade. In its acoustics, as in its light, the ocean grows cathedral-like.

In the gathering hush of the open ocean, humpbacks are the choir. They are spread out over thousands of miles of migration route, and their voices come intermittently. Migrating humpbacks do not pause for the continuous, daylong, one-whale symphonies they will sing on their wintering grounds, but they do sing, and their voices carry.

With a steady beat of his great piebald flukes, the humpback sails through the shafts of light; alone, yet surrounded by the voices of his brethren. They are all singing the same song, but chiming in at different times. Now and again, in the crowded solitude of the open ocean, the young bull adds his own voice to that ancient, ever changing round.

For the humpback, the open ocean is an old familiar road. For a human diver, it is alien.

Open-ocean diving is a peculiar experience. You hang from the surface by your snorkel, rising and falling with the rhythm of the swell. Beneath your fins lies a mile or two of ocean—a bottomlessness, an infinity of blue. You feel the pull of the void. The blue emptiness wants you, tries to call you down. A thousand beams of sunlight probe it, illuminating nothing. There is nothing to illuminate. Nothing provides any sense of distance or scale. A fish egg, or a tiny fragment of jellyfish, or a bit of salp drifting up ten inches from your faceplate becomes, for an instant, anything: a human, a whale, a fast-closing mako shark.

"Mark," wrote Herman Melville, "how when sailors in a dead calm bathe in the open sea—mark how closely they hug their ship and only coast along her sides." Melville knew why. "The awful lonesomeness is intolerable. The intense concentration of self in the middle of such a heartless immensity, my God! who can tell it?"

After a few days adrift in a small boat, fear fades a bit. Boredom sets in. On the boat or in the water, time crawls by as in a sensory-deprivation experiment. Hours pass, and sometimes days, in which you see nothing at all. The open ocean appears to be a biological desert. The long blue swells are sunny, sparkling, beautiful—and empty. The water is crystalline, like desert air. If anything were moving near the surface, it would be visible. Nothing is. Nothing exists but a deep, featureless blue.

Eternities of varying duration pass, and then something appears. As often as not, that something is a ctenophore. Glancing over the side, you see a gauzy, wraithlike shape—sometimes winged, sometimes ellipsoidal—passing underneath the boat.

The comb jellies of the phylum Ctenophora are made like jellyfish, their design just as translucent and delicate. Ctenophores are hardly there at all. They are, despite this immateriality, lethal predators on larval fish and little crustaceans. Where jellyfish and other cnidarians are armed

with nematocysts—small coiled stingers that pierce the victim's skin— ctenophores are armed with adhesive colloblasts, or "lasso" cells. Instead of poisoning their prey with thousands of little hypodermics, they rope it with thousands of tiny lariats.

The name ctenophore means "comb-bearer." Ctenophores move not by a pulsing of bells, like jellyfish, but by a beating of comblike rows of cilia that lie along their meridians. The facets of the moving cilia diffract sunlight. Tiny lights wink along the eight meridians—violets, reds, greens, blues. The pulsing lights are more intense, more definitely *there* than is the transparent protoplasm generating them.

Powered by rainbow-drive, insubstantial as a dream, the ctenophore passes from view. The desert of the ocean is empty again. The shafts of sunlight probe the blue, encountering nothing. The ctenophore might have been a mirage.

The next creatures to make their appearance, after the passage of another eternity or two, might be amberjacks. The fish are suddenly there. You never see them arriving. They hang, violently beautiful, in the fluid webs of light just under the surface. "Amber" does not do justice by them. The amber of their individual scales is edged exquisitely in pale salmon. A dark, diagonal bar runs through the eye, calling attention to the bright, ruby-amber iris. The amberjacks have swum in from the warm end of the spectrum. After epochs adrift at the cool blue end, you want to weep with gratitude.

The open ocean seems barren, it's true. Compared to the polychrome complexity of the coral reef; or Antarctic waters in their December bloom; or the cold, nutrient-rich currents that upwell along the coasts of Peru or southern Africa, the open ocean is a desert. Yet that oceanic desert, like the terrestrial kind, is secretly and variously alive.

Here the herds of antelope are schools of yellowfin and bluefin tuna. The cougars are solitary, quarter-ton marlin. The vultures are frigate-birds, great soarers patrolling the outer margins of the desert on black, motionless wings spanning seven feet. Below the frigates, the ocean's hawks—boobies, shearwaters, gannets, petrels, albatrosses, tropicbirds— hunt the ceaselessly marching dunes of the swells. Their prey, instead of mice and rabbits, are the flyingfish, needlefish, and assorted small fry that hide in the wrinkles of the surface.

Among the creatures of either desert, cryptic coloration is the rule. In the open sea, among the larger epipelagic animals—birds, billfish, tuna, sharks, rays, dolphins, whales—the camouflage colors almost make a uniform. These creatures are countershaded. The top half is dark, the bottom half pale. Seen from above, the creature blends in with the fathomless blue beneath it. Seen from below, it blends with the bright sky above. Smaller fish like herring, in addition to their countershading, have a silvery band along either flank. Backing the silver is black pigment, as in a mirror, and the flanks make mirrors indeed. Where the predator comes looking for *shape*, the mirrors show it the whole ocean. Reflecting the ambient blue, they cause the fish to vanish from most fields of view.

Among the minuscule animals of the plankton—crabs, krill, salps, larval fish, cephalopods, and copepods—the strategy is exactly the opposite. Most of these tiny drifters are transparent or translucent. They achieve

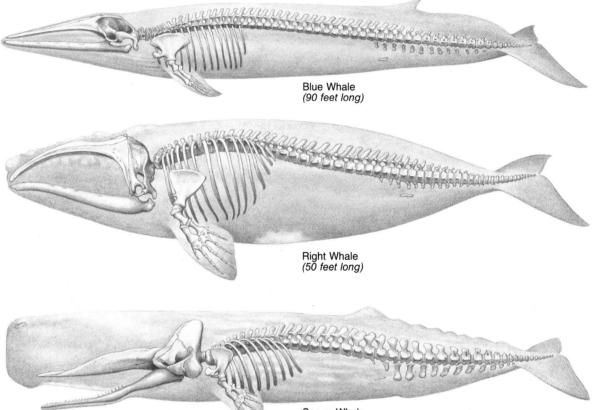

Blue Whale
(90 feet long)

Right Whale
(50 feet long)

Sperm Whale
(60 feet long)

invisibility by reflecting nothing at all.

When invisibility fails them, many fish take refuge in numbers. The predator, on solving all the mirror tricks and countershading, on detecting shape at last, is confronted by *duplication* of shapes. This is one function of the school. Prey fish, by schooling, appear to confuse predators, forcing a decision among a multitude of possibilities. Still another open-ocean refuge is in speed. The epipelagic zone of sunlight, with few places to hide, is the realm of the ocean's fastest sprinters.

The most streamlined of bodies, in water or air, are fusiform—spindle-shaped, tapering at either end. The uppermost ocean is dominated by fusiform fish. To penetrate this last refuge, to chase the runners down, top predators must be superfish. The tuna are examples. Of fusiform fish, they are among the roundest in cross-section, and thus closest in design to aircraft fuselages. Tuna are made like blue silvery bombs. Rotund at midpoint, they narrow to a wasp-waistedness in the tail stalk. Their rigid tail fins are capable of 15 beats per second, a remarkable rate for fish of their size. Bluefin tuna have been clocked at 40 knots. Tuna are so built for speed that in slow turns they are unstable. They avoid this embarrassment by seldom traveling slowly. Each bomb-shaped fish is heavy with muscle, some of it red, for endurance and high-speed cruising, some of it white, for the finishing kick.

In the desert of the open sea there are oases.

Gulfweed, or sargassum, makes one such place. Clumps of the weed support varied life in the exceedingly still, warm, salty, nutrient-poor waters of the Atlantic's Sargasso Sea. Without the weed's small, floating islands of golden-brown, the Sargasso would be among the most barren seas on earth.

Sargassum is an alga. Its small amber flotation bladders reminded Portuguese sailors of *sargaço*, little grapes, and thus the name. Seen from below—the viewpoint of the young turtles, tuna, marlin, swordfish, dolphinfish, and other pelagic species that shelter here—sargassum is remarkably like a grape arbor. Light streams through the fronds and their vintage of flotation bulbs, about the size of young Thompson seedless. Clumps of it, Columbus's journal reported, helped reassure his uneasy crew that land was nearby. "At dawn they saw many more weeds, apparently river weeds, and among them a live crab, which the Admiral kept, and says that these are sure signs of land, being never found eighty leagues out at sea."

The Sargasso Sea is bounded by currents: the Gulf Stream to the west, the North Atlantic Current to the north, the Canary Current to the east, the North Equatorial Current to the south. They flow into one another to make the strongest of all ocean gyres. The Sargasso Sea is the calm "eye" inside. The currents, and the winds that generate them, tend to confine the weed to the eye. It occurs there generally in scattered, pillow-size clumps. Occasionally the clumps are blown together in undulant rafts of an acre or two, but they never mass thickly enough to entangle a vessel—in spite of the persistent myth that they do.

For more than three centuries after Columbus, it was assumed that the weeds had detached from some shore or submerged shelf. Today the

Since they hang nearly weightless in water, whales do not need heavy skeletons for support. Spongy and porous, whale bones comprise only about 15 percent of the whale's body weight. In contrast, a land mammal's skeleton makes up at least half its weight.

The whale's flukes power its swimming. Its tail vertebrae are well developed; they support muscles that connect the flukes to nearly half the rear body.

Streamlined to move efficiently through water, whales have compressed necks, smooth skin, and a layer of blubber covering their bodies. Only their flippers, movable at the shoulders, protrude.

A whale's way of life is reflected in its body shape: The long and slender blue whale swims more quickly than the blimp-shaped right whale; the sturdy sperm whale, able to withstand changes of pressure and temperature, is well suited to deep diving.

consensus is that most sargassum is pelagic. It belongs on the open sea and spends its whole existence there.

Most of the fauna of sargassum are descended from bottom-dwelling shore species. Sargassum animals are desperately loyal to their weed. They adopt its shapes and colors; the crabs camouflaging themselves in mottled sargassic hues, the pipefish aping the undulations of the fronds, the sargassumfish sprouting leafy protuberances that mimic their namesake. Sargassum animals seldom risk the swim between one clump and the next. That reluctance, along with sargassum's method of propagation, makes for species compositions that are highly variable. Sargassum grows by partitioning. Whatever creatures happen to be occupying a portion of sargassum when it breaks free become the ancestors of the fauna of the new sargassum clump. Each community is a historical accident.

The Pacific has its own artificial sargassum. Anything that bobs on the open ocean—drift logs, fish floats, scrub brushes, brandy bottles, the lids to tennis cans—soon attracts colonists. There is almost always a crab or two riding atop each piece of flotsam. Small fish cluster below in a tight ball. Orbiting at a distance are the predators: blue runners, Pacific kingfish, sharks. Like clumps of sargassum, each piece of flotsam has its individuality. Each makes its own world, as different as Earth from Mars.

If flotsam and sargassum make one sort of marine oasis, another has its springs in upwellings near large oceanic islands. The upwellings reverse "nutrient trapping"—the tendency of minerals and organic matter to sink below the surface zones of sunlight, where phytoplankton can make use of them. The upwellings bring the fertile water back up to the light.

The clear, blue water leeward of the Hawaiian Islands is that kind of oasis. It is the piece of open ocean I myself know best. I spent six months of my life there, adrift in a small boat, doing stories on dolphins, whales, and other inhabitants. The blue water is a strange life. Three days go by, or four, in which you see nothing at all. You go half crazy. When, finally, on the fifth day, or on the sixth, the ocean relents and you see something, it is often a thing few humans have glimpsed before, or no human ever. That's wonderful. But you have paid for it with hours of no sights. You have paid in sunburned shoulders, cracked lips, and fits of moodiness.

On good days, the ocean and pure chance sent cetaceans my way—five species of small dolphins, sperm whales, beaked whales, even a Blainville's beaked whale. A Blainville's beaked whale, like other beaked whales, is an improbable sort of creation. It has a whale-size body joined to a small dolphinlike head. It was the first I had heard of such a creature.

The open ocean seems trackless to the eye, but in fact is all crisscrossed with highways. The lanes and mileposts are geomagnetic, olfactory, acoustic, celestial. The migratory traffic they handle is vast.

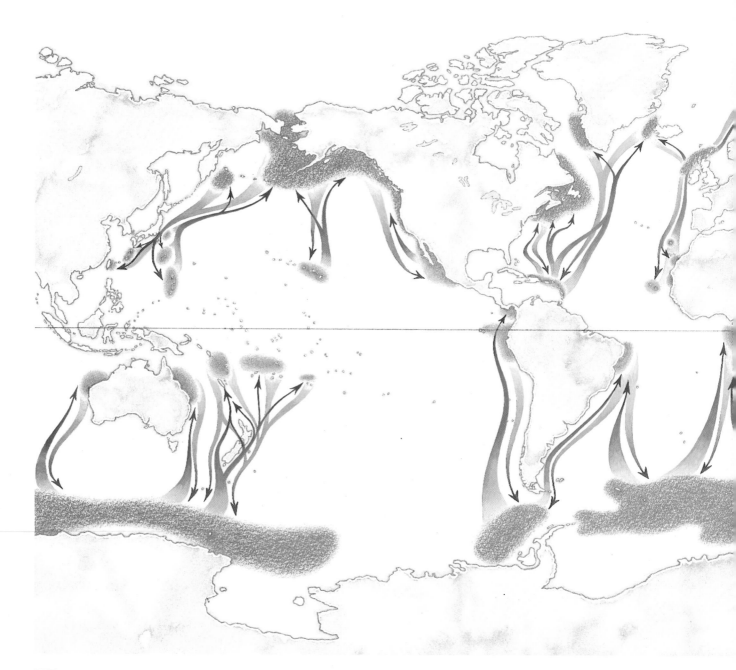

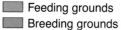
Feeding grounds
Breeding grounds

The arrows show only where the whales begin and end their journeys, not their actual migration routes.

Whether in the Southern or the Northern Hemisphere, humpback whales migrate north between March and May (blue arrows) and south between October *and December (red arrows). Individuals from the two hemispheres probably never mingle.*

In the eastern Pacific, for example, while one *stock of southern humpbacks breeds near Central America, their northern counterparts feed near Alaska; when the northern humpbacks travel to*

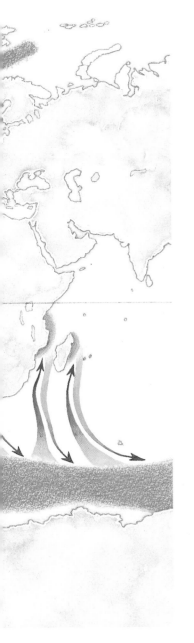

Hawaii to breed, the southern ones return to their Antarctic feeding grounds. Scientists still have much to learn about whale navigation.

Some migrations are anadromous, as of salmon and shad; heroic journeys from salt water back to natal freshwater streams to spawn. Some migrations are catadromous, journeys back the other way, as of the freshwater eels that travel from streams in Europe and North America across thousands of salty miles to spawning grounds in the Sargasso Sea. Many migrations are entirely pelagic, and scientists have traced the routes only sketchily, or not at all.

The salmon seem to find their way home by smell. The shad, giants of the herring family, are guided by smell and by temperature changes in the water. The marlin have magnetite in their heads and navigate, perhaps, by earth's magnetic fields. Arctic terns, among the greatest migrators of all, appear to steer by the stars.

For most of these epics, we are familiar only with the homestretch. We see the migrants only at their finish lines. But these are, after all, the climaxes, and there are few more moving events in nature: salmon leaping up their waterfalls or sea turtles returning to nest on their natal beaches years after leaving as hatchlings, the beach's bearings and coordinates somehow imprinted in the faithful reptilian brain.

These great horizontal migrations are what catch the human imagination, yet none of them is so momentous or vast as the daily vertical migrations triggered by darkness, and then by light. There are seasons in the sea, but no marine cycle matches the rising of creatures at dusk and their settling again at the glow of dawn.

The invention of echo-sounding devices provided the first pictures of vertical migration, though no one realized it at first. The captains of ships equipped with the new devices, in passing across stretches of open ocean, were often alarmed by echoes from unmarked shoals. Not wanting to linger for a survey, they hurried nervously on, but afterward dutifully reported the shoal to a hydrographic office. Later ships, passing the same spot, would report nothing but blue water, miles deep; but the phantom shoals found their way onto the charts. They were often named after wives and girlfriends of the sailors who had discovered them—"Betty Reef" and "Margaret Reef," for instance—and were generally marked ED, "Existence Doubtful," on the map. In print they have had an embarrassing persistence, like those same "Bettys" and "Margarets" tattooed on the sailors' chests.

In 1942, physicists experimenting with underwater sound off San Diego, working toward a method for echolocation of enemy submarines, were troubled by unexpected back talk. Their pings kept returning from a mysterious reflecting layer 900 feet down. The layer did not send back crisp echoes, as the hull of a submarine would have done; the echoes were soft and diffused over wide tracts of ocean, as if rebounding

from a false bottom. The layer was seldom so thick that the echo of the real bottom could not be heard through it. The stratum was named "Deep-Scattering Layer," or DSL, for its sound-scattering properties. This was amended to DSLs, as it became apparent that there were often three separate layers, and sometimes as many as five.

The San Diego physicists first thought the cause was physical, some thermal discontinuity in the water, but no one could imagine why such a discontinuity would migrate nightly to the surface. When biologist Martin Johnson of the Scripps Institution of Oceanography heard of the DSL, he almost instantly deduced the correct answer: the false bottom was alive.

Two groups predominate in the deep-scattering layers: the crustaceans, primarily euphausiids and sergestids, and fish, especially the lanternfish. The crustaceans are by far the more numerous, but the lanternfish are larger, and many lanternfish species have swim bladders, which make them far better sound reflectors.

The deserts of the open ocean, like deserts of land, come alive at night. When darkness falls on the Mojave or the Gobi, the desert creatures come out. When it falls on the Pacific or Atlantic, they come up.

In terrestrial deserts, the exodus of small creatures from their crevices and burrows is followed by predators with big, dark-adapted eyes—bobcats, coyotes, dingos, owls. In the sea, the same big, dark-adapted eyes prowl upward in the sleek bodies of squid and various midwater fish. In the dry desert, pit vipers follow heat trails laid down by kangaroo mice and hares, and desert cats pivot their great, pointed ears toward imperceptible sounds. In the wet desert, sharks follow scent trails laid down by their prey. They turn toward low-frequency vibrations detected by sensors in their lateral lines. They investigate electrical fields detected by their ampullae, the sensory organs that dot their snouts.

In either desert, day is the time for great vistas, night the time to meet the animals.

Most of the mystery of the open ocean, we tend to think, resides in its deeps. Men have been going down to the sea in boats for millennia; a thousand fleets have plied the surface—time and opportunity enough, it would seem, to clear up most superficial secrets. In fact there are enduring mysteries at all levels of the ocean. The top is nearly the *mare incognitum* that the bottom is. If we have recently discovered an unexpected dynamism in the abyss, then the same is true on the surface of the sea.

The top millimeter of the ocean is proving to be a crucial layer. It is the skin through which the ocean transpires and the pasture on which a disproportionate share of marine photosynthesis occurs. The surface makes a strange submicroscopic scenery. There are forests of dry surfactants, oily surface molecules with long hydrocarbon tails projecting into the air.

127

There are swamps of wet surfactants, proteins and glycoproteins with most of their hydrocarbon chains submerged. The top millimeter has its own biota, the neuston, tiny plants and animals specialized for existence on the air-ocean interface.

The wind and the neuston collaborate to help the ocean breathe. The smallest breeze-generated ripples increase by fourfold the transfer of carbon dioxide across the surface. (Their effect is in thinning the boundary layer across which the gas diffuses.) Waves contribute another twofold increase. The neuston, by stirring the boundary layer with their whiplike flagella, can increase evaporation and gas transfer by a factor of three.

We are only now discovering the importance of viruses in the sea. In marine ecology, as in physics, the hunt has been for ever smaller particles. In the 1960s and 1970s, marine ecologists concentrated their researches on nanoplanktonic algae. These tiny plants, less than 10 to 20 microns in size, seemed to be the ocean's most important primary producers. In the 1980s, attention shifted to picoplankton, bacteria less than 2 microns in size, which were proving more numerous in the seas than anyone had dreamed. Today attention has shifted to femtoplankton, viruses of less than 0.2 microns. A single teaspoon of seawater commonly contains 75 million viruses—many thousands of times more than previously estimated.

Our textbooks have yet to catch up. The simple linear food chains we have been taught in biology classes for most of this century—phytoplankton to copepods to fish to whales—are not the whole story. The food pyramids we memorized—algae at the base, whales at the apex—have to be modified. The marine food chain is not linear but labyrinthine, a web at whose base bacteria, protozoa, and viruses may play a large part. It is hard to say what sits on top or on the bottom, because a virus, or enough of them, can devastate a tuna or a whale. Fish and whales, it turns out, are footnotes in the story of the sea. Most biomass in the ocean, most energy flow, most carbon fixation, is in and by organisms too small to see.

Not all the unknowns of the open ocean lie at the minuscule end of the scale. Three months before the discovery of the hydrothermal communities in the Galápagos Rift, the upper ocean gave up a secret less momentous, probably, but nearly as unexpected and strange.

In the morning of November 15, 1976, the *AFB-14,* a Navy research vessel working 25 miles northeast of Oahu in waters two and a half miles deep, deployed two large nylon parachutes as sea anchors. For hours the parachutes hung at a depth of 540 feet. In the afternoon, when the chutes were winched up again, a shark more than 14 feet long and weighing 1,653 pounds was entangled in one of them.

The shark was outlandish. It had an enormous mouth, a great thick

129

tongue, and large protrusible jaws and lips. Nothing remotely resembling it had been seen before. It represented a new species, genus, and family. It was novel down to the very gut. In the valvular intestine, parasitologists found tapeworms of a genus and species undescribed. Scientists and reporters called the shark "megamouth," and the taxonomists went along with that, naming it *Megachasma pelagios.*

Megamouth is a fine symbol for the ocean's unknown. It is type species for all the unspecified, dimmer-than-shadowy orders and families and genera it has left behind. It represents all those creatures that from rareness, or speed, or intelligence, or pure chance have escaped the trawls of science and have yet to surface in human awareness.

Someplace off the South Sandwich Islands, in May, after three years' ocean wandering, a female right whale feels the stirrings of an old urge.

The migration routes of right whales, *Eubalaena glacialis australis,* are still poorly known, but a bit is understood of their reproductive cycles. For the southern right whale, the calving interval is about three years. This female's last calf has gone its own way. Heavy feeding on the copepods of the Southern Hemisphere summer have made her plump again. Southern winter is approaching. She finds herself heading north for the breeding grounds off Argentina. She does not make good time. Right whales are the portliest of cetaceans. Their slowness, and the quantities of oil and baleen they yielded, and their tendency to float after death, all combined to make them the "right" whales to pursue in the days of longboat whaling.

It is hard, despite their sad history, not to see right whales as the most jovial of big cetaceans. There is the right-whale girthiness. There is the right-whale curiosity—a fatal trait in the days of Basque and Yankee whalers. There is the outlandish and comic face. The right whale's eye lies low, at the corner of its mouth. From there the jawline makes a steep curve up and forward, so that most of the mouth rides high above the eyes. The visage is wonderfully homely, covered everywhere with lumpy white callosities, and at the same time wonderfully homey, for those callosities are densely inhabited by whale lice.

And there is the right-whale exuberance. Despite their weight, or because of it, right whales leap energetically. Breaching in cetaceans has been correlated with rotundity. The slowest, roundest whales—gray whales, humpbacks, right whales—breach most often. Cetologists have suggested that breaching may sometimes be playing, other times a show of aggression. Breaching may be a way of signaling identity. Whether different species and kinds of jumps communicate different meanings is still unknown. Perhaps the humpback, in twisting its 40 tons five-sixths out of

The grand courtship dance of the wandering albatross results in a single large egg that is incubated for 80 days. This largest of all seabirds breeds in high-latitude areas, such as Albatross Island off the coast of South Georgia (above), where strong wind gusts can support takeoff.

water—huge flippers inscribing crazy arcs, spray flying—makes a boom on crashing back that proclaims "Humpback!"

The right-whale cow, loafing north for Argentina, is a last vestige of her race. The southern right whale's prewhaling population has been estimated at about 100,000. Of that number, only two or three thousand remain. This whale has little to be cheerful about, it would seem, yet suddenly now she accelerates. Swimming under the surface and parallel to it, she develops speed that is only fair, but momentum that is colossal. She raises flukes, tilts her head up, breaches, crashes back. *"Right whale!"* she signals, to the sea or sky, or to distant fellows, or to herself.

Isolated by water, life thrives on many patches of earth that poke above the waves. Even on an active volcano like Fernandina in the Galápagos Islands (below), which rises out of the sea from earth's crust, dense greenery flourishes alongside new ash.

Random drifters from far continents accidentally populate oceanic islands. The rigors of ocean travel filter out most species; successful immigrants evolve independently.

Marine iguanas, the world's only seagoing lizards (opposite and pages 134–135), presumably floated to the Galápagos from South America, where they probably originated millions of years ago as terrestrial animals. They gradually developed the appurtenances of sea life: a laterally flattened tail for propelling themselves in water; a blunt snout and sharp teeth for gnawing algae off rocks; glands for distilling excess salt from their bloodstreams (to be ejected nasally); and scaly black skin for resisting ultraviolet sun rays.

Opposite and Pages 134–135:
Amblyrhynchus cristatus, Marine Iguana, 39 inches long

Octopus, 5 inches long

Alectis ciliaris, African Pompano

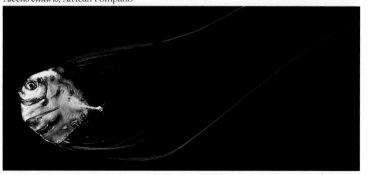

Scientists study the Galápagos and Hawaiian archipelagoes as laboratories of evolution. The Galápagos total roughly half the area of the Hawaiian Islands and are about 65 million years younger. They lie more than 1,500 miles closer to continents, and thus to sources of terrestrial life. Hawaii's eight main islands in the North Pacific Ocean are the most remote on earth. Early settlers named Kahoolawe (far left) for the Hawaiian god of the ocean. A dot on the map south of Maui, Kahoolawe receives less than ten inches of rain per year, making much of it desert.

In the open ocean, as in terrestrial deserts, many animals emerge only at night. One such creature photographed near Hawaii, a three-inch baby African pompano (lower left), trails 15-inch filaments, which could pass for jellyfish tentacles. The tiny young octopus at left contracts and expands special pigment sacs called

Monachus schauinslandi, Hawaiian Monk Seal, 7 feet long

chromatophores to change color. A transparent bulb reveals and protects its internal organs. Transparency works as camouflage and may also resist harmful ultraviolet light.

Hawaii and the Galápagos both have their own mammals. Galápagos sea lions (left) glide and dance through the water in harems, each dominated by a bull weighing 600 pounds or more. They mate underwater, perhaps to avoid overheating. Some 50,000 sea lions populate the Galápagos, whereas in Hawaii only about 1,400 of their monk seal cousins survive. The endangered monk seal (above) is the oldest and most primitive of living seals, unchanged for 15 million years. Annually in spring, females weighing up to 600 pounds lumber ashore to give birth to their pups. Over a six-week period, the mother lives off her stored blubber while nourishing her pup, which grows from its birth weight of 35 pounds to about 140 pounds.

Every two or three years the green sea turtle performs an astounding navigational feat: It travels a thousand miles or more between feeding and breeding grounds. The turtle's routes are a mystery, but recent research suggests it keeps on course by sensing and responding to wave motion.

When not mating, these turtles, weighing about 350 pounds each, are dispersed throughout the warm waters of the Atlantic, Pacific, and Indian Oceans. But it is only near Australia that they thrive in anything like their natural abundance. Humans have threatened the green turtles' existence since the 16th century, killing millions for meat, soup, oil, and shells, and reducing the population to near extinction in most places.

Now protected by many countries, the turtles are making a comeback.

Green turtles are primarily herbivores, so the creole fish at right have little to fear. Toothless, the turtles rely on horny, serrated jaw sheaths for their diet of algae and sea grasses. Occasionally they do eat animal matter, especially if it is clinging to the vegetation on which they happen to be dining.

The best feeding areas are far from the native nesting beaches to which the turtles return to breed. Strong and graceful, they swim submerged, holding their breaths for long periods. But in spite of their strength, once there they must rest before facing the arduous tasks of breeding and egg laying.

Males are smaller and may be more numerous

Pages 140–145: *Chelonia mydas,* Green Sea Turtle, about 40 inches long

than females. They compete vigorously for female favors. Males may wait in line, but often one or more will try to dislodge a successful competitor (top), which will hang onto the female even if wounded in the battle.

Mating takes place in the offshore shallows, the male clutching the female with his long tail and front flipper claws. He sometimes clings so tightly that his claws cut notches in the front edge of the female's shell. During mating, both partners must somehow keep their heads above water to breathe. Copulation may go on for six hours.

Then, under cover of darkness, at or near high tide, the massive and bulky female strains up onto the beach, past the high water mark. She digs a nest about 15 inches deep that will accommodate her body. Under her tail, she digs out another, smaller hole some 15 inches deeper. She lays about a hundred soft, white eggs bathed in mucus (above). Then she covers the nest with sand (opposite),

burying the eggs deep enough to keep the temperature constant and the sand moist, but shallow enough so hatchlings will not suffocate before they dig out. Finally, her entire mission accomplished within three hours, she drags herself back to sea, leaving a distinctive trail behind (opposite). The female shows no further interest in her future offspring, but repeats her egg-laying ritual up to eight times during the season.

After about 60 days, the hatchlings emerge from their nests at night and scurry for the sea. Having eluded crabs and birds on shore, the hatchling above has heightened its chances of survival.

Perhaps to keep its mother in view, a newborn humpback calf swims above and to one side of her head, but whale calves are curious and often venture away from their stations to explore. Though rarely have humans witnessed a whale birth in the wild, scientists generally believe that calves are born both during migration and in wintering grounds such as Hawaii. There, a mother and calf, lit by rays of sun that penetrate the ocean surface, swim near shore (left).

By the time it is one year old, the calf eats plankton and small schooling fish, and is ready to fend for itself. The cow now begins to push her calf away.

Humpbacks attain sexual maturity when they reach 39 feet or so in length, but continue growing until they are about 50 feet long. Both males and females mate repeatedly with different partners throughout the six-month breeding season.

Courting males are aggressive, and they sing. Their songs are complex, with themes repeated precisely by all adult males in the same winter breeding area. While singing, a humpback male stays away from the rest of the pod. The hopeful suitor positions himself less than 150 feet below the ocean surface, stretches out his flippers, and, with his head down, inclines his body 45 degrees. In what might be a more direct stimulation to mate, a humpback male blows bubbles that gently

At left and Pages 148–149: *Megaptera novaeangliae,* Humpback Whale

rise toward the female's genitals (opposite).

As she nears the end of her 10- to 12-month pregnancy, the humpback cow searches out a shallow inshore haven where she can escape bull whales and sharks. Once her 2-ton, 15-foot calf is born, it quickly figures out how to swim, breathe, and feed on its mother's milk. Mother and calf remain inshore for about a week, then venture into deeper waters two or three miles out.

A filter feeder, the humpback is outfitted with baleen, comblike plates fringed with intertwining bristles that hang from its upper jaw. The baleen is made of keratin, the protein in hair, nails, and claws. To feed, the whale swims through schooling fish or krill with its mouth open (above). Elastic ligaments permit its jaws to

open wide. It gulps a great mouthful of fish and seawater, then spends a few moments expelling water through the baleen, which traps the food like a sieve.

Another filter feeder, the sleek fin whale (right), swims with energy and grace. Weighing about 80 tons, it closely resembles the blue whale, but never reaches its size. The fin whale travels the world, though less in tropical than in temperate and polar waters. It even ventures into inshore passages, as well as the Mediterranean Sea and the Gulf of California. Fin whales sometimes swim singly or in pairs, but usually move in pods of three to seven. An individual can dive 775 feet or deeper. On surfacing, it powerfully ejects from its

blowhole a 13- to 20-foot jet of vapor shaped like an inverted pyramid.

The blue whale (page 152, top), largest of all creatures, also glides through the Gulf of California. At 100 tons and 90 feet long, it requires about 4 tons of krill a day to satisfy its appetite. The blue whale is fast for its size, normally swimming about 14 miles an hour.

By contrast, the right whale (page 152, bottom) is a slow swimmer, averaging about 2.5 to 3.5 miles per hour. It migrates with the seasons, but follows less regular paths than humpbacks, bowheads, or gray whales.

Balaenoptera musculus, Blue Whale

Eubalaena glacialis australis, Southern Right Whale, 50 feet long

The 16-foot pilot whale (opposite), along with bats and other toothed whales, can emit sound pulses that reflect off objects, creating echoes. Called echolocation, this sonarlike ability enables the pilot whale to construct acoustic pictures of its surroundings and thus locate the squid, other cephalopods, and small fish it preys on.

Unique among whales, the gray whale (pages 154–155) finds its food on the seafloor. It dives down, rolls onto its side, and sucks in sediment. Using its tongue to force out water and silt through its baleen, it retains huge quantities of amphipods— crustaceans a third to an inch long.

Temperate Seas

The right whale, making the deliberate progress of her species, navigating by means unknown to humans, steams for her breeding ground off Patagonia. She passes above the submarine elevations of the Falkland Plateau, then above the depths of the Argentine Abyssal Plain. The cold sea grows warmer. A storm comes up, and she dallies, breaching and lobtailing in mountainous waves. The tumult of breaking seas seems to excite right whales, to call forth all their ponderous acrobatics. They are moved to contribute some tumult of their own.

The storm subsides, and the whale resumes her migration. Beneath her the seafloor ascends, sloping gently upward in the continental rise, steepening in the continental slope, then leveling in the continental shelf of South America. The clarity of the open ocean gives way to the cloudier waters of the Argentine coast. The whale is surrounded now by convivial sounds: the breaching, head slapping, lobtailing, and odd moans, grunts, and dyspeptic trumpet notes of other migrating right whales. She raises the headlands of Peninsula Valdés. Her long journey done, she enters Golfo Nuevo.

She is greeted by a buzz of dolphin click-trains.

Ahead, in the greenish water of the gulf, a school of anchovies takes shape. The school darkens, flashes silvery white, then darkens again as the frantic fish dart this way and that. The anchovies are now a ball, now a bolus, now a fleeing wall of fish. Streaking in and out of visibility, orbiting the school at high speed, are the dusky dolphins that are shaping it. The sea pops and sizzles with dolphin click-trains, and the fish flinch, turn, scatter, and tighten in a cross fire of sound.

Dusky dolphins are a nearshore species native to all southern temperate seas. They are among the most acrobatic and streamlined of dolphins. Their flanks are painted in racing stripes, their beaks hydrodynamically whittled down. Now and then a dusky dolphin breaks off its herding to slash through the ball, and anchovies disappear as if by magic.

The whale enters the dolphin-anchovy maelstrom—or the maelstrom

Filtered sunlight illuminates an underwater ballet of Australian sea lions in the temperate seas off South Australia. Once aggressively hunted, these rare and graceful pinnipeds are now protected from humans, while still a dietary staple of great white sharks. Though clumsy ashore, sea lions can rove miles inland.

careens into the whale. For several instants the sea is all confusion: anchovies reversing contrasts as they change direction, dolphins streaking from below, black-browed albatrosses and cormorants diving from above. Silvery contrails of bubbles mark the paths of the birds, and the surface is a bright turbulence of foam and webbed feet. Then, as quickly as it came, the maelstrom passes. The hunt moves on. The water is green and empty again, except for the silvery scales of the fish.

The whale rests a while, loafing at the surface. Five right-whale bulls arrive, a welcoming committee. They bump each other in reasonably friendly competition, hundreds of tons of amorousness colliding as they jostle for position underneath the female. She twists away, but the bulls pursue her. She should be more cooperative, probably, given the low numbers of her kind, but she is not in the mood. Holding her breath, she rolls on her back, presenting her piebald underside to the sky. When the cow is in this posture, there is not much the bulls can do with her.

One bull holds his breath, rolls, and stations himself belly-up underneath the female. When she runs out of air, she must roll back to bring her blowhole out of water. Embracing her sides delicately with his great rounded flippers, he awaits this great moment. The battle of the sexes becomes a breath-holding contest.

The cow has tricks of her own. After 20 minutes on her back, she finally runs short of air. Surging ahead, she quickly blows, inhales, and then stands on her head in the water, her flukes in the sky. To align with her properly, the bull must lift his own flukes out of water, upon which he loses all maneuverability. She takes the opportunity to nap. The earnestness of her suitors, the plight of her race, will have to wait. She dozes, unmolestable, sailing gently downwind as her flukes catch the breeze.

The Peninsula Valdés, the southern right whale's breeding ground, is one of the planet's more remarkable meetings of land and sea. The comings together here seem impossibly exotic: from land, armadillos and the American camels called guanacos and the American ostriches called Darwin's rheas; from sea, right whales, orcas, dusky, Risso's, and bottlenose dolphins, southern sea lions, Magellanic penguins.

But all such meetings of land and ocean are remarkable: The comings together of southeast Alaska, for example—grizzly bear and salmon, wolf and northern sea lion—or the comings together of New York's Gateway National Recreation Area, where the students of Public School 145 meet the horseshoe crab.

Temperate seas are the seas many of us know best, as schoolchildren, scientists, fishermen, poets.

Our first marine lessons were in the intertidal zone, where the ocean drops regularly and conveniently to let us explore. Intertidal communities are most extensive in temperate latitudes. Antarctica has none, because of its ice cap. In the Arctic, tidal amplitude is small, and the zone between tide lines is frozen for most of the year. In much of the tropics, tidal amplitude is low as well. In the temperate zone, strong tides and moderate climate have led to a flowering of intertidal life.

Rocky temperate shores are among the sea's richest environments. They provide a solid substratum for attachment by plants and animals, strong wave action, clean water. It is along rocky shores that the zones of the intertidal area are most clearly delineated. The husband-and-wife team of T. A. and Anne Stephenson, British pioneers in rocky-shore ecology, staked out three zones in that narrow but lengthy province of life. On his Pacific shores their American counterpart, Ed Ricketts, co-author of *Between Pacific Tides,* and hero, lightly fictionalized, of John Steinbeck's *Cannery Row,* detected four zones.

The highest zone the Stephensons called the Supralittoral Fringe. In other schemes it is called the splash zone or the littorina zone. Ricketts called it uppermost horizon. By whatever name, it belongs more to land than to sea. Lying above all but the highest tides, its top is wet only by ocean spray and storm waves. It is the realm of periwinkles, limpets, rock lice, barnacles. Its bottom margins are patrolled by rock crabs, its upper borders colonized by lichens, its middle elevations darkened by the microscopic plants of the "black band," which runs clear around the planet's temperate shores, like a ring left by a somewhat higher ocean. The band is composed in small part by lichens and green algae, in large part by blue-green algae, a life-form nearly as old as the sea itself.

The greatest stress of the supralittoral is desiccation. As protection against it, the blue-green algae of the black band are encased in mucilaginous sheaths. The periwinkles—marine snails of genus *Littorina*, shore dweller—have mastered several strategies against it. When the tide falls and the sun dries the rocks, a periwinkle seeks out the shade inside cracks and under ledges. Withdrawing into its house, it attaches itself by mucous threads to the rock, then seals the horny door of its operculum. In Latin, *operculum* means "lid." Rough on the outside, pearly on the inside, the operculum shuts out desiccating air while shutting in seawater. Behind their opercula, periwinkles can survive high and dry for more than a month, and can tolerate immersion in fresh water for several days.

Ed Ricketts once gave the operculum a tough, if accidental, test. To keep a captive anemone alive, he fed it a periwinkle, assuming that the caustic digestive juices would batter down the operculum door. Some hours later, when the anemone disgorged it, the periwinkle emerged alive, like Jonah. Its shell was beautifully cleaned and polished. On

159

In the cool, nutrient-rich waters of Monterey Bay, a diver surveys one of the world's largest submarine forests, composed of giant kelp. The coarse seaweed's fronds, kept afloat by gas-filled bladders, may grow a hundred feet in a single season—as much as one foot a day. Strong and supple stemlike stipes carry sugar alcohols, the product of photosynthesis, from the kelp's sunlit upper blades to dimly lit sections of the plant below. Secured to underwater rocks by branched holdfasts, kelp shelter a vast and complex community of fish, sea otters, and many other marine animals and plants.

A large tide pool on Moresby Island, in British Columbia, is home to a community of rapacious sea stars, which prey on barnacles, mussels, and other shellfish. Aquariums of seashore life, tide pools form on rocky shores wherever hollows, crevices, or natural dams trap seawater. Twice a day the tides resupply the pool with fresh seawater and food. Every tide pool has its own distinctive plant and animal population. This may include the stalked solitary tunicate (opposite, top), which attaches head-down to a rock; and the sandcastle worm, whose tentacles snare plankton from the incoming tides (opposite, bottom).

cautiously opening its operculum, it praised God, then went gliding off toward Nineveh.

An evolutionary drama is under way among the periwinkles. They are transitional animals, sea creatures in the process of becoming terrestrial. The stages in this metamorphosis are apparent in the snails of various zones. In the three New England species, for example, the smooth periwinkle, which dwells lowest in the intertidal, can survive only briefly in air. The common periwinkle, dwelling higher, spends more time in air than water, yet still must lay its eggs in the sea. The rough periwinkle, dwelling highest, is viviparous—gives birth to live young—and possesses a gill cavity that works almost like a lung. Where the lower-living periwinkles excrete metabolic waste as urea, the rough periwinkle excretes it as uric acid, which is insoluble and saves water in the elimination.

Out of old habit, the rough periwinkle is most active at two-week intervals, with the high spring tides. It seems, however, that almost anytime, with a little courage, the rough periwinkle could shrug off that vestigial loyalty and head for the interior.

A similar drama is under way among the rock lice. These isopods, genus *Ligia,* are those little sea pill bugs one sees scurrying on beach rocks. They are like the pill bugs of land, only faster and equipped with a pair of long, spiny appendages behind. Like terrestrial pill bugs, they are gregarious. One sees dozens or hundreds, seldom just a few. Their habituation to land seems very far advanced. When the tide is out, *Ligia* wanders down into the intertidal. When the tide turns, it beats a retreat.

"*Ligia* is very careful to avoid wetting its feet," Ed Ricketts wrote. This is a telling observation, revealing of *Ligia* and its commitment to land, and revealing of Ricketts himself. Ricketts personally hated getting his *head* wet, if we are to believe John Steinbeck. ("He will wade in a tide pool up to the chest without feeling damp, but a drop of rain water on his head makes him panicky.") In the little rock louse, the great intertidalist seemed to have recognized a kindred spirit.

The next zone, the Midlittoral Zone, is the section exposed and covered daily by the high tides. In other schemes it is called the high intertidal or the balanoid (barnacle) zone. Conditions in this middle region are not so harsh as in the zone above. The midlittoral is a richer place, both in species diversity and in numbers of individuals. This is the zone of the brown seaweeds. The indicator species are the rockweeds, brown algae of genus *Fucus;* and the acorn barnacles, genus *Balanus,* which often form a band at the top of the zone; and the mussels, genus *Mytilus,* which form their beds somewhat lower. There are also crabs, chitons, sea stars, goose barnacles.

The midlittoral has an underclass. Beneath loosely embedded rocks, protected there from surf, sun, and desiccation, lives a fauna characterized by its squirminess and friability: brittle stars, which shed their writhing arms at the slightest provocation; ribbon worms, which break into sections on contact; slithery, clinging fish called blennies.

The last zone, the Infralittoral Fringe, is the section uncovered only by very low tides. These low-intertidal rocks belong almost entirely to the sea. The infralittoral fringe has greater species diversity than the other intertidal zones. Almost all the earth's plant and animal phyla have representatives here. Species are so numerous and competitive that there is not room for great single-species armies like those of the periwinkles and barnacles. This is the zone of the red seaweeds. In protected places, every inch of rock is forested in the reds, pinks, and purples of the nether rockweeds and coralline or laminarian algae. "Triton is a fruitful deity, and here, if anywhere, is his shrine," wrote Ricketts.

Tide pools occur in all three zones. The highest pools are sparsely inhabited, the lowest densely. The pools in between are richer places than their middling position would suggest. These middle tide pools, in approximating the conditions of the rich infralittoral fringe below, become high outposts of that zone. The anemones, hydroids, and sponges characteristic of the low intertidal are able to live there, and can be seen without waiting for low spring tides. For humans, tide pools are a great convenience, and they made the first marine laboratories.

Below the last tide pool, in the subtidal rocky bottoms along many temperate shores, grows the kelp forest. The large kelp are the redwoods of the macroalgae. Off the Pacific coast of South America grow kelp whose stipes, or stalks, stretch 200 feet from their holdfasts on the seafloor to the surface, where they spread their canopy of fronds. Kelp off the coast of California can grow 100 feet in a single season. The fronds are capable of adding one foot of new growth each day.

Kelp forests can be more productive than tropical rain forests. "The number of living creatures of all Orders, whose existence intimately depends on the kelp, is wonderful," wrote Charles Darwin. "A great volume might be written, describing the inhabitants of one of these beds of seaweed." The kelp forest understory is often as lavishly multicolored as a coral reef and nearly as diverse. Anemones, sponges, and corals encrust the bottom. The holdfasts by which the kelp attach to rocks on the seafloor are habitats for clams, small crabs, snails, tube worms, brittle stars, sea urchins. Higher up, amidst the trunks of the stipes in the middle stories of the forest, bright orange garibaldis hold position against the current. Higher still, in the canopy, the kelp fronds are host to epiphytic algae—tiny cousins of kelp—and to epiphytic animals such as hydroids, cnidarians, bryozoans.

Intertidal life in sand is more difficult, sparse, and secretive than life on the rocks.

Beaches are continually reshaping themselves. Their profiles change with the seasons. The sand grains are small and in continual flux. Sandy shores provide no reliable footing for attachment. The holdfast of kelp, the tight fit of limpet, the muscular foot of abalone, the calcareous fortress of barnacle, the mucous adhesive of periwinkle, the strong byssal threads anchoring the mussel, are all worthless in sand. Where creatures of the rocky intertidal defeat wave shock by attachment, creatures of the sand defeat it by burying themselves. To escape surf, sun, desiccation, and enemies, sand creatures dig down.

Visitors to the beach come in various sizes—sandpipers hunting, turtles nesting, humans combing, the occasional whale beaching itself. The permanent inhabitants come in just two sizes. There is a macrofauna of the familiar clam-size, crab-size burrowers; and there is the meiofauna, creatures of a millimeter or less dwelling between the grains of sand.

Among the macroscopic burrowers there are several strategies. Some are fast, deep diggers, like the razor clam. The razor slices down rapidly to depths where the sand is stable. Others are fast, shallow diggers, like the coquina clam. On being washed out, the coquina quickly reburies itself a half inch deep. In a series of burials and resurrections, stealing the energy of the surf, the coquina works its way up the beach on flood tide, down the beach on ebb. A few are slow, shallow diggers, like the Pismo clam. Pismo clams are creatures of violent surf, and they *require* that violence, dying within days if moved to calmer waters. They hold position in their rioting waters by the heaviness of their shells.

One of the more beautiful adaptations to this simple, shifting universe is the mole crab.

The mole crab really is remarkably molelike—the same general shape, the small, dim eyes, the sensory whiskers, the flattened pawlike appendages, the burrowing instinct, the industriousness. The crab's shell is egg-shaped to distribute the surf's pressure evenly, allowing the crab to keep its balance in the tumult. Underneath, the crab is equipped like one of the more expensive Swiss Army knives. It comes jammed with an assortment of retractable tools: stalked eyes, sensory bristles, a pair of short antennae for breathing, a pair of long antennae for filter feeding, bottlebrush appendages for cleaning the filters, paddlelike hind appendages for swimming, legs for crawling and burrowing, all of them folding together neatly inside the oval of the shell. The crab is at home anywhere in the surf zone, whether swimming above the sand, crawling on top of it, or digging below. Wherever it goes—in whichever of these modes—the mole crab travels backward.

166

Mole crabs display the gregariousness of many intertidal species. For reasons unknown, they feed shoulder to shoulder in great armies. Their technique is to settle tail first into the sand until only their stalked eyes and breathing antennae show above. They face the surf in tight phalanxes, like the legions of Caligula in his crazy war on the sea. They make no move to feed in incoming waves. They sit tight until each wave has spent itself high on the beach, then wait out the undulations of its backwash. When the wave has thinned to a returning sheet of water, the mole soldiers, as if on signal, simultaneously uncoil their feeding antennae. Each crab waggles the feathery fork of its antennae in victory V at the sea.

The antennae are wonderfully efficient nets. They catch dinoflagellates, plant particles, and perhaps even bacteria, netting particles from four to five thousandths of a millimeter to two millimeters in size. The mole crab does not strain at gnats and swallow camels. It strains and swallows gnats and camels with equal ease. If the macrofauna of the sandy shore has produced an ultimate animal, perhaps it is this sand mole.

The creatures of the meiofauna are invisible, or nearly so, yet in the sands they are vastly more important than creatures we can see. Rachel Carson described the meiofauna beautifully: "Lilliputian beings swimming through dark pools that lie between the grains of sand."

Swimming Carson's dark interstitial pools, or inhabiting the film of ocean that makes a minor water planet of each wet grain, are small polychaete worms, flatworms, nematodes, rotifers, foraminiferans, arthropod mites, harpactacoid copepods, gastrotrichs, tardigrades. Theirs is a creation at least as strange as the one Gulliver stumbled across. Even the names are Swiftian.

The tardigrade, or water bear, is the teddy of the meiofauna. No more endearing animal ever crawled into the radiance of the electron microscope. Tardigrades are roly-poly and lumbering—in their microscopic way. The small phylum Tardigrada is primitive, lying midway between the gastrotrichs and the arthropods. The creatures are equal parts mite and earthworm, yet in body plan—externally, at least—they really do look eerily bearlike. They might be miniature bears reduced, by magic spell or wrong evolutionary turn, to microscopic size.

The water bears are bearlike even in practicing a kind of hibernation. When dried out, they shrivel, suspend their animation, and enter what biologists call a cryptobiotic state. Immersed in pure alcohol, or in strong and desiccant salt brines, or in liquid helium at −488°F, they endure in some interior compartment of themselves and return to life on being moistened. Tardigrades may be the ultimate adaptation to the rigors and extremes of the sandy shore. If the beach meiofauna has produced an ultimate animal, perhaps it is this sand bear.

Salt marshes form in protected areas along coasts, where sediments from rivers or the sea can accumulate. Though the marshes may look quiet, their profusion of grasses and other plants, seeds, and insects feeds an enormous and varied assortment of fish and other animals; myriad birds such as the great egret (right) eat the fish. Salt marshes benefit humans too, since they also nurture most of the seafood we consume.

Behind a barrier of dunes, along many of the temperate ocean's sandier shores, lies the salt marsh.

In this marsh the old, elemental struggle between land and sea has ended in a truce. The marsh belongs to both worlds equally. Skies are huge over the salt marsh, horizons are long, the wind in the spartina grass has the sound of a prairie wind, and a visitor might imagine himself on the High Plains—until he hears the chitinous click and scrape, the softly thunderous rustle, of the herds moving through the grass. The buffalo here are fiddler crabs. The stampedes follow the tides. The smells are of grass, yes, but also of iodine and of the good muck of the marsh, a sweet-salt richness of decay.

Land and sea make a fertile interface, along most temperate shores, but nowhere more so than here. Professor Eugene Odum, the marsh's great explainer, first motored up a Georgia tidal creek on an ebb tide in 1954. He was struck by the golden hue of the mudbanks to either side. He marveled at the six-foot stands of marsh grass topping the banks—grass as luxuriant, he thought, as a well-fertilized stand of sugarcane. "The notion came to us, in those early days, that we were in the arteries of a remarkable energy-absorbing natural system whose heart was the pumping action of the tides," he has written.

The mud, Professor Odum and his students were to learn, owed its

In Atlantic and Gulf coast estuaries, when spring tides reach their highest levels, horseshoe crabs come out to mate. At the water's edge, females dig holes in the sand and lay eggs, which the males then fertilize. Dangers—storms, seabirds, and especially humans, who harvest them for bait and medical research—imperil the crabs from infancy to adulthood. Even so, these unusual creatures, which are not actually crabs but arthropods (and thus more akin to spiders), have survived on earth for 360 million years.

Pages 170–171: A territorial dispute between two American lobsters will end in death for the loser. Overfishing off New England has devastated the population of these scavenging crustaceans, but biologists estimate that the species can endure if only 2 of every 10,000 spawned reach adulthood.

golden color to huge numbers of diatoms and other algae resident in the banks. His impression that the marsh was like a well-fertilized field of cane turned out to be short of the mark. Further studies showed Odum that where a field of Hawaiian sugarcane annually produced 14.8 tons of dry weight per acre, and where a midwestern cornfield produced 5.7 tons, the salt marsh produced 17.8.

Odum's wild notion that the tidal creeks were arteries in some sort of superorganism proved entirely correct. The entire salt-marsh complex of creeks, marshes, river mouths, and barrier islands is a single community linked by the tides, its fecundity powered both by the rays of the sun and by the tidal pull of the moon.

At first glance, the salt marsh seems to be all one species, *Spartina alterniflora,* salt-marsh cordgrass. The marsh, on the face of it, is one of those monocultures that nature is supposed to abhor. At second glance, with aid of microscope, the variety becomes manifest. The marsh is revealed as a teeming polycultural jungle of hundreds of species of diatoms, green algae, blue-green algae, and dinoflagellates.

The mud and its flora are a nutrient trap for phosphorus, calcium, and other elements vital to life. The marsh grass itself, when it crumbles to detritus, feeds the microorganisms that feed larval fish, larval shrimp, mussels, oysters, periwinkles, whimbrels, turnstones, plovers, clapper rails, blackbirds, ospreys, ibises, egrets, herons, river otters, raccoons.

"If a quarter acre of marsh could be lifted up and shaken in the air," writes John McPhee, "anchovies would fall out, and crabs, menhaden, croakers, butterfish, flounders, tonguefish, squid. Bigger things eat the things that eat the marsh, and thus the marsh is the broad base of a marine-food pyramid that ultimately breaks the surface to feed the appetite of man."

Seaward of its inshore habitats, the sea bottom slopes down into the plain of the continental shelf, and the salt marsh, kelp forest, and seagrass bed give way to benthic sand and mud. It is on the continental shelf that our most important fisheries lie—for shrimp, lobsters, crabs, clams, cod, sardines, rockfish, bottom fish.

The shelf sediments are inhabited by an infauna of burrowers such as polychaete worms, clams, and tunneling crustaceans, and an epifauna of anemones, brittle stars, sea urchins, sea stars, ambulatory crustaceans, and other creatures adapted to life on the surface of the sediments. Nutrients arrive on the shelves from two directions—in runoff from land and in upwelling along the continental margins—and they fuel plankton blooms over the shelves. While plankton photosynthesize above, diatoms inhabiting the shelf mud photosynthesize below. Organic detritus

The mottled coloration of a goosefish off the Maine coast allows it to lie in wait for prey by blending in with the sandy bottom. Protruding from behind its upper lip, a fleshy "fishing lure" twitches, drawing curious fish, which are instantly snapped up in the goosefish's huge, toothy mouth. Inshore cousins to deep-sea anglerfish, the voracious goosefish (also known as monkfish) devours not only squid, lobsters, and other fish, but diving seabirds, too.

Another coastal bottom-dweller and disguise artist, the windowpane flounder (opposite) has its color pattern and both eyes on the left side of its body. The fish is as adept as a chameleon at varying its color to match the surroundings.

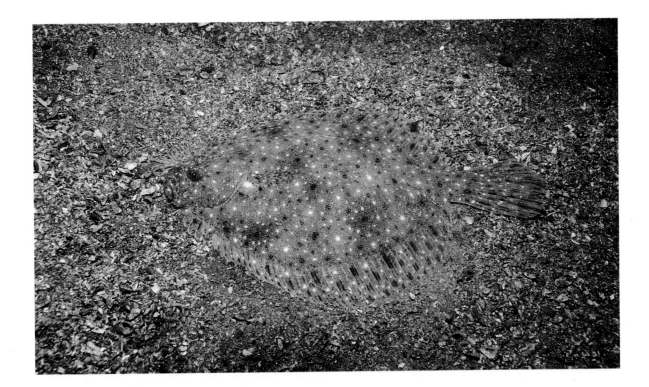

Pages 174–175: Outfitted in winter plumage, a dense flock of migrating dowitchers and western sandpipers rests along the Washington coast. In fall and winter, the high-pitched jeets and keeeks of these shorebirds are common on beaches, mud flats, and tidal creeks on both the Pacific and Atlantic coasts as far south as South America.

outwelling from the salt marsh, kelp forest, and sea-grass pasture is harvested by filter feeders that extract it from the water, and by deposit feeders that mine it from the mud. For humanity, all this makes for ecosystems both fecund and accessible.

If the shores and shelves of temperate seas were where humans began to understand the ocean, then these same shores and shelves are where, of late, we have begun to understand the damage we are doing there.

Summer of 1988 seems to have been the season of awakening. In the summer of 1988, more than a million bottom fish died in New Jersey's Raritan Bay, trapped in a dead zone of anoxic water. In the summer of 1988, some 18,000 seals died in the North Sea, killed by pneumonia and liver infection. In the summer of 1988, medical wastes began washing up on beaches from Long Island to northern New Jersey—bloody bandages, syringes, prescription bottles, vials of blood.

Chinook salmon caught nowadays off the coast of Oregon are affected with tumors. The fish and shellfish of New Bedford Harbor, Boston Harbor, Buzzards Bay, and Narragansett Bay are contaminated by high levels of toxins. The North Sea and the Baltic Sea are dying; large regions of them are anoxic and lifeless, killed by algal blooms. The blooms are stimulated by runoff of nitrogen and phosphorus from factories, sewage

plants, and farms. Baltic seals are starving, their mouths and flippers so deformed that they are unable to catch fish.

We have begun the attempt to make amends. Restoration has begun along the shelves and shores.

Biologists have learned a good deal about how to resurrect urban estuaries. Kelp forests have been replanted in southern California. China has established mangroves on its eroding southern shorelines. Sea grasses are being restored in the Philippines.

There have been successes, too, in the repopulation of marine creatures. With protection and reintroduction, the sea otter has slowly begun to reestablish itself on long stretches of the California and Alaska coasts. The gray whale, with protection, has rebounded from a low of only a few thousand to today's 21,000. Hunted to near oblivion by the end of the 1800s, the northern elephant seal has multiplied from less than 100 to today's 80,000 to 90,000.

The "great sweet mother," Swinburne called the ocean, and on that the poets and paleontologists agree. It's past time we began treating her so.

A pod of long-finned pilot whales, departing the depths of the Newfoundland basin, pass above the Grand Banks of Newfoundland's continental shelf and make for the shore.

The prey of *Globicephala malaena,* the long-finned pilot whale, is the short-finned squid. In summer, the squid migrate inshore to spawn, and the pilot whales follow them there. Occasionally, pursuing squid into shallow bays, the whales strand themselves. In prehistoric times, for beachcombing foxes, ravens, and humans, these mass strandings made an embarrassment of riches. In the minds of the humans, they may have sparked an idea. Why wait for the whales to beach themselves? Why not help the process along?

Newfoundland's pilot whales have had rough treatment at the hands of humans. The island drove a "fishery" for pilots, which, in the first three centuries of its existence, may have been sustainable. Newfoundlanders took just one or two thousand pilots annually. Then mink ranchers moved to the island, in search of cheap meat to feed their caged animals. They began converting pilot whales into mink coats. In 1956, the slaughter reached its peak—nearly 10,000 whales. As the population plummeted, the annual catch fell to hundreds, and finally, in 1971, to just six. In 1972 the fishery closed.

Today the numbers of pilot whales are rebounding. The whales are reclaiming their old territory, renewing their acquaintance with their traditional prey. In the murky green waters of those once-fatal shallows, their biosonar detects squid, and their peglike teeth quickly close on them.

175

CANADA'S STRAIT OF GEORGIA

An inland sea separating mainland British Columbia from Vancouver Island, the Strait of Georgia abounds with fjords, islands, and estuaries. Tidal currents surge through it, bringing in water so rich that a single quart was found to hold millions of minute plants and animals.

This planktonic feast attracts many larger crea-tures, which, in turn, draw fishermen and scuba divers in large numbers. Because the plankton is so dense in summer, scuba divers prefer the frigid, low-light days of winter (below), when underwater visibility can extend a hundred feet.

What awaits them is a wealth of animals as colorful as those of the best tropical reefs. A giant octopus, the world's largest at five feet long, trails eight sucker-covered arms past several sea anemones (opposite). A master at hiding and hunting, the shy creature propels itself by jetting water through a siphon tube. It lives under rocks, darting out to stun a crab with paralyzing venom before devouring it.

Left: *Octopus dolfeini,* Giant Octopus

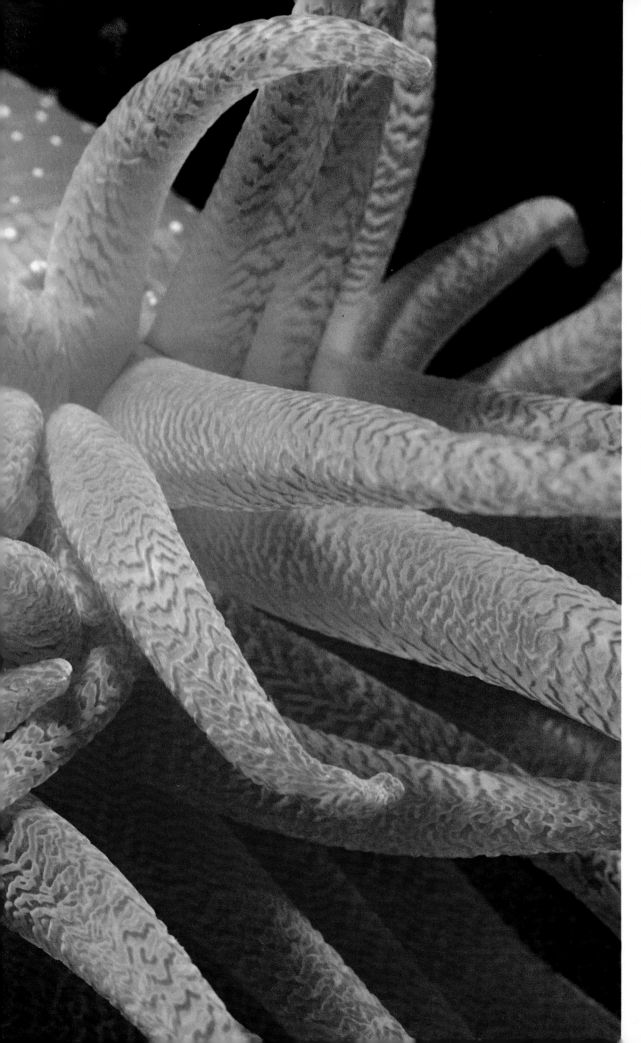

The column of a textured pink anemone is home to a clown shrimp (left), which feeds on its host's wastes. To avoid attracting the anemone's attention, the shrimp slathers itself in the larger creature's mucus.

Pages 180–181: Three communities of orcas, also known as killer whales, inhabit the Strait of Georgia and nearby waters. Divided into close-knit pods, these social and intelligent animals (which are actually a type of dolphin) may gain up to 6 tons on a diet mostly of fish. In contrast to the resident orcas, transient orcas hunt down and devour other whales, dolphins, and seals, but no orca has ever been known to kill a human. Each pod has its own repertoire of sounds, consisting of whistles, screams, and grunts.

Cribrinopsis fernaldi, Anemone; *Lebbeus grandimanas,* Clown Shrimp, 1 inch long

Pages 180–181: *Orcinus orca,* Orca, 30 feet long

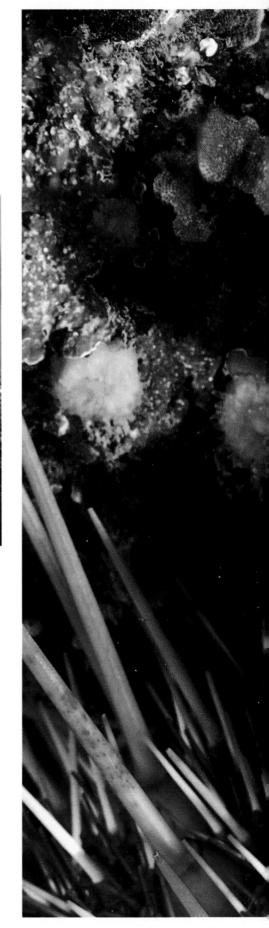

Anarrhichthys ocellatus, Wolf Eel

Dark lines running out from its eyes like tears identify the penpoint gunnel, an uncommon eel-like creature that inhabits the shallows, often in pairs. Here, off an island in the Strait of Georgia, a gunnel snakes its body around a red sea urchin (right). Colored according to diet and habitat, this gunnel's greenish tinge blends well with eelgrass; red or brown gunnels live among rocks. The name "penpoint" comes from the large penlike spine that protrudes from the gunnel's anus.

Another elongated creature resembling an eel, the six-foot-long wolf eel (above) is a giant cousin of gunnels and blennies. The wolf eel here makes short work of a spiny sea urchin; it is one of few fish with the dental equipment to do so. Inside powerful jaws, the wolf eel has stout canine teeth in front and double sets of molars in back—a "pavement of molars," says one scientist.

Though the doughy-faced wolf eel may look like a monster, it is actually shy and reclusive. With its lifelong mate, it spends

Chlamys hastata, Pacific Pink Scallop, 2 inches

years in a single den.

A nearly circular Pacific pink scallop (above) extracts microscopic food particles from seawater by rhythmically pumping the water through its two-part ribbed shell, called a bivalve. At the shell's edges glow some of the mollusc's 30 to 40 blue-green eyes; each has a cornea, lens, and retina, but can detect only light, not shapes or color. To swim freely or escape from danger, the scallop claps its valves together and forces water out through an opening near the hinge, in a kind of jet propulsion maneuver.

Preying on its own kind, an orange Dawson's sun star (right) appears to be moving in on a group of ochre sea stars, which eat mussels, limpets, snails, and barnacles. Common to the Pacific Northwest's rocky intertidal areas, both star species gather food by means of their hundreds of tube feet and mobile stomachs. The stomachs can be pushed out and spread over prey and even inserted into a shell.

Solaster dawsoni, Dawson's Sun Star, 15 inches; *Pisaster ochraceus,* Ochre Sea Star, 10 inches

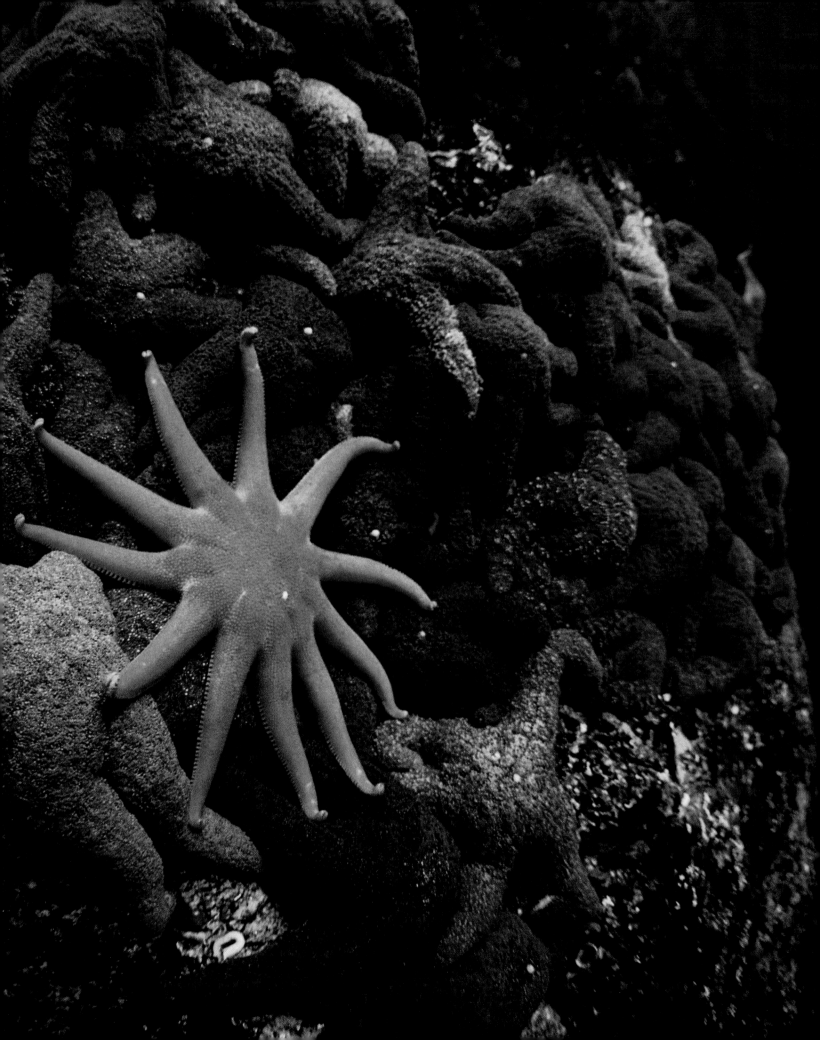

M onterey Bay cuts a watery, nutrient-filled crescent into the central California coast, feeding a submarine kelp forest and the creatures that dwell in it. Nutrients are replenished seasonally by upwellings of cold waters laden with organic and nonorganic matter and by mineral runoff from the land. So much food is available that many marine species are limited mainly by places to cling, crawl, or burrow.

In the kelp-latticed waters of the bay, invertebrates resembling corals, anemones, and sponges vie for space on a rock wall (left). Among the invertebrates, the red coral-like *Corynactis californica* wards off predators with knobby tentacles laden with stinging capsules.

Common in Monterey waters, the six-inch-high white-spotted anemone (below), named for the spots on its column, clings to rocks and feeds on whatever brushes by, including crabs, jellyfish, and sea stars.

A tiny red rock shrimp ventures right up to a California moray eel's mouth

Tealia lofotensis, White-spotted Anemone

Opposite: *Corynactis californica,* Strawberry Anemone; *Lagenipora* sp., Bryozoan Colony; *Hippodiplosia insculpta,* Bryozoan Colony; Tunicate

Opposite: *Gymnothorax mordax,* California Moray, up to 5 feet long

to feed (opposite), risking its life. The shrimp's meal consists of dead tissue and parasites that have accumulated on the eel's skin. The eel, in turn, eats shrimp and other crustaceans, octopuses, and small fish, preferring the night hours for hunting and feeding. With its long, slender body, it is able to follow an octopus into a hiding place and devour it. Usually this involves agile maneuvering to disengage tentacles the octopus has fastened to the eel's head.

The crevice kelpfish (above) protects itself by closely matching its color to its dwelling place. Whenever it changes location, it may take up to a week adjusting to the new color. The coralline algae in which this kelpfish nestles thrive in temperate waters.

Pages 190–191: Deep in the inshore kelp forest, a young harbor seal weaves through cool strands of the seaweed. Distinguished by its small size, chunky proportions, small front flippers, and spots, the harbor seal ranges

Pages 190–191: *Phoca vitulina,* Harbor Seal, 5 feet long; *Macrocystis pyrifera,* Giant Kelp, 100 feet tall

Above: *Gibbonsia montereyensis,* Crevice Kelpfish, up to 4 inches long; Coralline Algae

Peprilus simillimus, Pacific Pompano, up to 11 inches long

widely and dives deep, hundreds of feet on occasion. While underwater, its body supplies blood only to vital organs—brain, heart, and lungs—by constricting peripheral vessels. To conserve energy, the seal reduces its heart rate from about 85 beats per minute to 15 or 20, and can even sleep while submerged for up to 6 minutes at a stretch. The purpose of these extended dives is to find food, cover long distances efficiently, and escape predators.

The jellyfish (right) is a visitor to the kelp forest. A simple creature, its mouth and stomach hang from the center of its mantle, and tentacles armed with stinging cells radiate outward to paralyze and then capture prey. But the tentacles can also serve as shelter for small fish, such as the juvenile Pacific pompano above.

South Australia's fertile waters churn with the energy of blizzards that originate thousands of miles away in Antarctica. In reclaiming the edge of the continent, the roiling waves have sculptured the Twelve Apostles (above), limestone pillars that stand off Port Campbell, southeast of Spencer Gulf.

Dangerous Reef, at the mouth of the gulf, is home to the great white shark (left). The most aggressive of sharks, the great white may be drawn to this region because of the presence of sea lions on which it feeds. It also eats tuna, dolphins, and other sharks, and scavenges carrion and garbage. Humans are not part of the shark's normal diet. Scientists suggest that the rare assaults may occur because of mistaken identity, invasion of shark territory, or bad timing: approaching a female with young or a male ready to mate.

Though little is known about great whites, scientists believe they are highly sensitive to smells, sounds, movement, and the earth's magnetic field;

Sepioloidea lineolata, Pinstripe Squid, 3 inches long

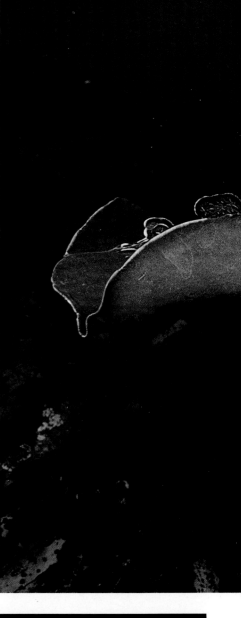

Hapalochlaena maculosa, Blue-ringed Octopus, 4 inches long

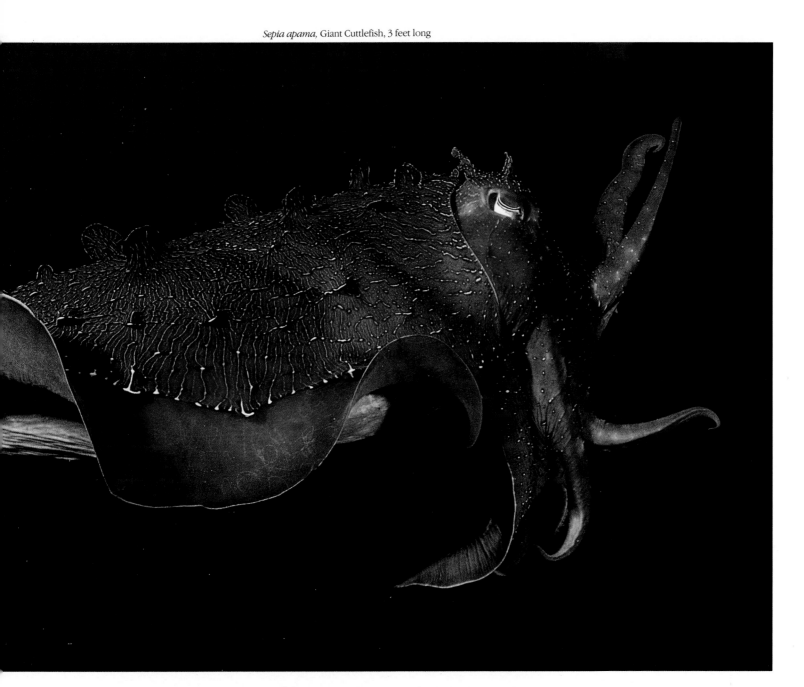

electromagnetic sensors in their snouts apparently help them pinpoint prey. How many of the sharks exist, how often they breed, how long they live, and whether they migrate are still mysteries. Observers have learned that they are large—probably up to 21 feet long and more than 7,000 pounds—mobile, and swift. They can cruise long distances at about 2 miles per hour, accelerating up to 15 miles per hour to attack.

The great whites, other sharks, and dolphins eat the giant cuttlefish that live along South Australia's rocky shores. The cuttlefish flails its arms to warn off intruders (above). It eats shrimp and crabs, using the same basic hunting strategy as the pinstripe squid (opposite, top). Each stalks its victim, then seizes it with two long, sucker-tipped tentacles. Eight shorter arms lined with suckers hold the prey fast, while a sharp beak injects toxin stored in the salivary glands.

The blue-ringed octopus (opposite, bottom) has eight arms of equal length.

197

Pyura spinifera, Ascidian; *Cochleoceps* sp., Clingfish, 1 inch long

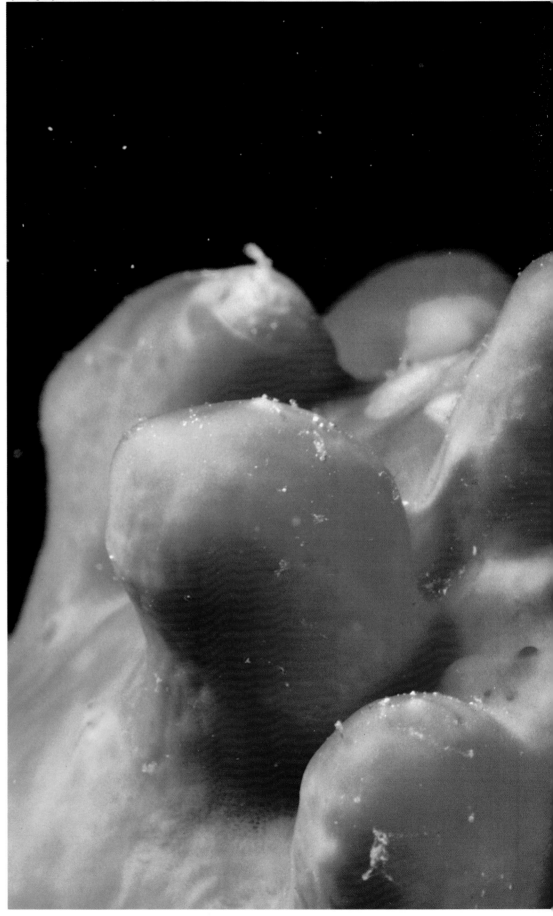

To trap a victim, it envelops the prey in its webbed arms or lashes out with an arm or two. When threatened, the tiny octopus flushes a vivid blue by expanding and contracting pigment sacs in its skin. If the warning is not heeded, the octopus may deliver a deadly bite.

The bumpy head of a brightly colored ascidian, or sea tulip, seen close up at right, shelters a tiny clingfish, camouflaged to avoid predators. The ascidian spends its first few hours swimming tadpole-like. Then it attaches itself to the seafloor, tail up. The tail dissolves, and the ascidian's body rotates so that the mouth points outward.

Pages 200–201: Black-faced cormorants station themselves on Dangerous Reef. From this perch, the eager fishermen can identify the tides and currents that bring the best catches.

Bordered on north, east, and west by the bountiful South Pacific and on the south by the harsh, polar Southern Ocean, New Zealand's temperate waters encircle thousands of miles of coastline. Off the west coast of South Island, where an early snow dusts Doubtful Sound (right), the bottlenose dolphins below cruise and hunt, their biosonar creating pictures of their surroundings. The dolphins feed on sea creatures, including squid and fish discarded by fishermen. In shallow waters, they often roll over and feed upside down, probably to enhance echolocation by subduing surface echoes.

At New Zealand's opposite end, the Poor Knights

Tursiops truncatus, Bottlenose Dolphin, 8 feet long

Parablennius laticlavius, Crested Blenny

Islands lie about 15 miles off North Island's northeast coast. These uninhabited bits of land overlook a magical underwater sanctuary where the sea has etched cavities out of volcanic rock. Among a throng of creatures, a pair of two-inch crested blennies (above) mate in a rock pocket close to shore. Nearby, a long-spined red urchin (right) feeds on seaweed and defends itself with two sets of mobile spines. Each spine, tucked into a ball-and-socket joint, has a muscle at its base. Between the spines, minuscule organs reflect light rays onto the photosensitive skin. Whenever a shadow flits over the urchin, its spines swivel in defense.

Diadema palmeri, Long-spined Red Urchin

Phocarctos hookeri, Hooker's Sea Lion, 7.5 feet long

Hippocampus abdominalis, Sea Horse, 7.5 inches long

The Hooker's sea lion (left) claims as its territory New Zealand's outermost southern islands, especially the Auckland Islands. Only 5,000 to 7,000 sea lions live and breed in the area—far fewer than in 1806 when the Aucklands were discovered, but many more than a century ago when a law ended sealing. The sea lions begin breeding in early December, with each male establishing a territory on the beach to receive a harem of females. Pups are born a year later.

A seven-and-a-half-inch sea horse (above) in a shallow bay on Stewart Island is probably a male. Its bloated abdomen may be carrying eggs laid by a female. Sea horse reproduction occurs when the female injects her eggs into the male's incubating pouch. He inseminates them, nurtures them, and after a few weeks spews them into the sea. It may take several days for as many as 300 half-inch babies to emerge. Hungry when born, sea horses eat crustacean larvae and other small creatures with their tubular mouths.

Another creature with an elongated snout, the southern pigfish (opposite) probes crevices for shellfish, crustaceans, and worms. An eight-to-ten-inch adult weighs about two pounds.

Clown nudibranchs (above), 1½-inch-long carnivorous sea slugs that feed on sponges, dwell off the Poor Knights Islands.

They possess both male and female organs, enabling them to play both sex roles. The one on the right is laying eggs.

Pages 210–211: In caves near the nudibranchs, jewel anemones reach out to trap plankton. Their deadly tentacles stun the microscopic creatures, then draw them to the anemones' waiting mouths.

Opposite: *Congiopodus leucopaecilus,* Pigfish

Pages 210–211: *Corynactis haddoni,* Jewel Anemone

Tropical Seas

pod of short-finned pilot whales, traveling in a purposeful, straight-line way and chirping like an aviary, follows its dominant bull toward the atoll of Ailinglapalap, in the Marshall Islands. At the narrow pass into the atoll's lagoon, the bull makes an odd choice. Instead of turning back toward the open ocean, he heads for the lagoon.

In the confines of the pass, the whales close ranks. The sandy floor beneath them is studded with coral heads, each populated by a swarm of brightly colored fish. As the whales pass, their great shadows extinguish the neon colors of the fish. Each darkened swarm tightens into the protection of the coral. The shadows pass, the neon colors rekindle, the swarms expand again.

The pilot whale is a large dolphin—the largest of all, after the killer whale. The dominant bull of this pod is 20 feet long. He has the bluff, bulbous forehead characteristic of old bulls, yet he is so big through the chest and shoulders that he seems pea-headed. His beak is insignificant. His mouth is fixed permanently in the odd, pursed smile of pilot whales. His caudal peduncle—the swelling of his tailstock just forward of the flukes—is covered, like the rest of him, with white scratch marks, deeper scars of combat, and circular half-dollar-size wounds left by cookie-cutter sharks. He looks every bit the leader, sea-tested and wise. The evidence shouts, though, that something is very wrong with him.

The sandy shallows of an atoll lagoon are the wrong place for pilot whales. Pilots are deep divers. Their staple food is squid. Nothing awaits the pilots here but trouble.

The mass stranding of toothed whales is one of the ocean's abiding mysteries. No whale is more subject to this apparently suicidal instinct for sand than the pilot. Long-finned pilot whales—the antitropical species—regularly strand on temperate shores, fleeing from or searching for God knows what. Short-finned pilots—the tropical species—are given to entering the atolls of the Pacific to die on the lagoon beaches in the equatorial sun. Autopsies of stranded whales sometimes show brain or sinus

*A moray eel slithers
through soft fan corals
140 feet down in Japan's
Izu Oceanic Park. With
mouth agape, it is ready
for a passing meal of fish
or crabs. The Kuroshio, the
western Pacific's equiva-
lent of the Gulf Stream,
warms Izu's otherwise
chilly waters, creating a
tropical marine sanctuary
only 60 miles from Tokyo.*

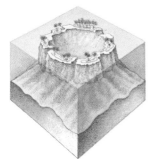

In tropical waters, three types of coral reef are associated with volcanic islands. A fringing reef (top) grows outward from an island coast. As the island subsides, its coast shrinks inward; the coral becomes a barrier reef, separated from the coast by a lagoon (middle). If the island sinks below sea level, an atoll (bottom), such as the Kayangel Atoll in the Pacific (opposite), remains.

damage from infestations of parasites, and it was thought at one time that the organisms damage the equilibrium, or the judgment, or the biosonar of the whales. The fact that baleen whales, which do not echolocate, never mass-strand, either, seems to point to a malfunction of biosonar. But there are other theories. Because stranding whales almost always come ashore on shelving sand beaches, some scientists have suggested that the problem is acoustic; that the shelving contour sends back mushy echoes that confuse the whales. The people of Ailinglapalap, for their part, believe the error is in judgment—that the whales are crazy.

If the whales inside the Ailinglapalap lagoon yet entertain any doubts about the sanity of their leader, they do not show it on their faces. Their mouths are fixed, as always, in the slightly foolish pilot-whale smile. The juvenile whales take the opportunity to explore the lagoon floor. Juvenile pilots are paler than their elders, less scratched and scarred, and, like juveniles everywhere, more curious.

One half-grown pilot glides close by a coral head and rolls to bring its eye to bear. A blenny, dark purple in its foreparts, Day-Glo orange in its tail, flees across the convolutions of a brain coral. A pair of Moorish idols sidle away from the whale. A yellow-faced angelfish shies into a crevice. It is an angelfish in *chador,* its lemon-yellow face hidden, below the level of the eyes, in the blue reticulations of a veil. The veiled angelfish looks more Moorish than the idols. It turns tail in the crevice to look out, then ducks back in again. Its body is marginated all around by fins trimmed in electric blue. The entire fish is haloed in that blue glow, as if struck just now by lightning, or martyred already and elevated to heaven.

In the life of the young whale, the most interesting pictures have been acoustic. The visual world has been mostly the ghostly light of bioluminescence by night and monochrome blue by day. The whale is not used to colors like these.

The coral reef is a paradox. It is an island of varied and unearthly color in the monotonous blue of the tropical sea. It is a rampart of intricate and infinitely various form against the formlessness of the open ocean. It is the richest of marine systems amidst warm surface waters that are in nutrients among the poorest on earth.

It begins when the small, transparent, ciliated coral larva—the planula—settles out of the plankton and attaches to a hard surface. Renouncing the motile life, the planula transforms into a polyp. It exchanges the cilia on which it swam for a central mouth surrounded by a ring of tentacles. The polyp extracts calcium and carbonate ions from the sea, building itself a skeleton, a snug, fortified cup of coral limestone. Each species of reef-building, or stony, coral constructs its fortress on a different

blueprint. The skeletons take the shape of stars, asterisks, craters, pores—hundreds and hundreds of configurations, each one distinctive.

The original settler divides asexually into 2 polyps, then into 4, then 16, eventually into 1,024, and so on. It grows into a colony, a cluster of genetically identical individual polyps. At this level of organization, too, each species builds after its own blueprint. The colony takes the shape of a plate, a branch, a table. There are button, bead, bubble, and brain corals; staghorn, vase, and organ-pipe corals.

For the purposes of nomenclature, science looks at the colony rather than the individual polyp. Corals are generally classified not for particular characteristics of the polyps themselves, but for the shapes of their skeletons, taken singly or en masse. Perhaps this seems a little unfair, but the polyp is an undistinguished little being, and the colony distinctive. It is understandable that science should have taken such a course in trying to make sense of the reef. Corals of the family Fungiidae, then, are named for their resemblance to fungi. The Tubiporidae, the tube-pores, are the organ-pipe corals. The Milleporidae, the thousand-pores, are the fire corals. The Siderastreidae are the starlet corals, and so on.

In the right circumstances—on shallow continental shelves in the tropics and subtropics, or on the shores of oceanic islands in the warm seas—the colonies aggregate to form reefs. The corals have help in this enterprise. If the stony corals are the building blocks of the reef, then calcareous algae are the cement. There are a host of secondary "frame fillers" as well—myriad small plants and animals whose calcareous shells decompose into the sand that fills in the reef's cavities.

At this level of organization, as at the level of polyp and colony, the coral community builds according to different blueprints. There are fewer of these, just three general types. Charles Darwin identified them: barrier reef, atoll, fringing reef. If the coral colony is a superorganism, then reefs are super-superorganisms. The barrier reef, most impressive of the three varieties, is a kind of long, gargantuan, biogeological *being* of limestone, chlorophyll, and stinging tentacles. The thousands of species of coral, alga, sponge, arthropod, mollusc, and fish that compose it together do bioengineering, as in their construction of the spur-and-groove system along the reef's outer edge. The spurs point seaward, toward the march of the swells. By deflecting incoming waves into the grooves, they force the sea to expend its energy against itself. The sea is forever trying to wash the reef away, but the reef outsmarts it.

The coral polyp is not much to look at—a circle of translucent tentacles, a mouth that doubles as anus—yet this creature is the greatest builder on earth. Coral reefs are the largest structures made by living things on the planet. The simple polyp fabricates—with help from its friends—an

Pages 216–217: *Their mild stings harmless, giant schools of amber* Mastigias *jellyfish swarm around a diver in one of Palau's marine lakes. The protected seawater lake in which the jellyfish live is one of about 80 in the western Pacific's Palau archipelago. Isolated from the surrounding ocean, each salt lake shelters a unique ecosystem, studied by scientists as a marine laboratory.*

Pages 220–221: *Palm trees grow right to the shoreline on one of the Russell Islands, part of the large Solomon group in the southwest Pacific. Coral reefs, anchored to the coast, begin growing where palms leave off.*

architecture that is huge, heterogeneous, and vastly complex.

For an eye new to it, the color and diversity of the reef can be overwhelming. There are moments of vertigo in trying to sort it all out. A solid wall of blue fusiliers parts ahead of you, darkening as the school turns in unison, reflecting the depths. The wall closes behind you, flashing silver as the fish, turning again, reflect the sun.

The outline of an octopus materializes in the sand. It moves hypnotically and double-gaited, jetting along with contractions of its siphon and rambling on its tentacles. It has an uncanny ability to assume the colors of its backdrop. Each time it stops, its existence becomes problematical.

Reaching some coral rubble, the octopus instantly assumes the blotchy purple of the coral. It stops there, or seems to have stopped, becoming a lump of rubble. Then comes a kind of after-image. You detect, or imagine you do, a ghostly movement to the side, as if the spirit of the octopus, in an out-of-body experience, has flowed onward into a cave in the rubble. Press your facemask close to the cave, and there indeed, staring back, is the goatlike satyr's eye of octopus. The octopus grows excited. Its chromatophores contract and expand in wild combinations, and colors race across its epidermis like fire.

In terrestrial landscapes, there are negative spaces where the eye finds rest. There is no rest for the eye on the reef. It is a seascape where every inch is adorned and busy.

In time, the eye begins to organize the reef's elements. Categories suggest themselves. There are the different strategies in use of color, for example. Some reef creatures, the sole, octopus, and stonefish among them, employ color as camouflage. Others, like the nudibranch, employ it in exactly the opposite way: to call attention to themselves. Many nudibranchs—sea slugs—have colors so garish they "set the teeth on edge," as one observer put it. That is exactly the idea. Nudibranchs feed by choice on toxic prey, such as sponges and sea anemones, rendering themselves poisonous. Color-coded less strikingly, nudibranchs would be eaten by mistake, to the considerable grief of all concerned.

A third group uses color in mimicry. The imitation cleaner wrasses, actually saber-toothed blennies, belong in this category. These small, duplicitous fish have adopted the color patterns and stylized dances of the cleaners. They set up phony cleaning stations, sting operations. On luring fish close to be cleaned, they bite a plug of flesh from the client. A fourth color strategy—the commonest of all, it seems—is that of those creatures that use color patterns as marks of their species and attractions for the opposite sex. The harlequin tuskfish and the clown triggerfish, two species painted as if for the circus, belong in this category.

"Aren't I pretty?" asks the harlequin tuskfish. "Beware," warns a nudibranch. "Try and find me," teases the sole. Many reef creatures make several suggestions simultaneously, of course, and the octopus is subject to trying them all.

It is a commonplace to describe the coral reef as a jungle. On a reef the very stones are textured in mouths—the tiny, tentacle-surrounded orifices of coral polyps. Larger jaws are everywhere: waiting in coral caves, hiding under coral sand, cruising in streamlined bodies through the reef's coral buttes and valleys. That the reef is a dangerous place is manifest in the reproductive strategy of its inhabitants. A large proportion of coral creatures spawn on outgoing tides, entrusting their eggs to the open ocean. The cruel sea is a safer place for larvae, apparently, than is the many-mouthed jungle of the reef.

The curious thing is that one so seldom witnesses violence on the reef. Predators abound, but one almost never sees predation. The crepuscular hours—the brief intervals of twilight at dawn and dusk—are feeding time on the reef. There is a tension then. The fish are faster and more jittery, and even the human diver gets caught up in the mood. If one is to see a strike or a capture, it is likely to be by twilight. But in full daylight there is little drama.

At night the reef takes on a different character. Many reef fish doze; some spin mucous sleeping cocoons for themselves, others seek out coral crevices. The corals themselves feed at night, it's true. The reef then has an almost floral beauty, if one shines artificial light on it. Flick the switch and there, all around, is a night-blooming of coral tentacles in a thousand hues. But much of the drama is microscopic, between plankton and polyp. The violence is all but invisible.

What one sees on the reef is not its antagonisms, but its alliances. The reef is a monument to collaboration: between cleanerfish and the fish they clean, between blind burrowing shrimp and the gobies that serve as their sentinels, between and among all the cementing algae, sea slugs, sea stars, feather worms, urchins, crustaceans, bivalves, fish, sea snakes, and sponges that have worked over millennia to build the reef.

At the very heart of the reef is the symbiosis between reef-building corals and their resident algae, the tiny one-celled plants called zooxanthellae. The corals derive food and oxygen from the zooxanthellae; the plants hide from predators and gain access to sunlight in the corals. Reefs whose corals contain the algae grow 90 percent faster than reefs whose corals do not, because the tiny plant cells speed up a coral's production of calcium carbonate, thereby adding skeleton to the reef. The reef's foundation, in the material world, is coral rock laid down by countless generations of coral polyps. Its foundation in the metaphysical world is the mutual dependency of coral and zooxanthellae.

Science came late to the coral reef. We have only begun to understand reef rhythms and processes. As many as half the species on the ocean's richest reefs are still unknown to science, according to some estimates. The unknowns are not just in specific details, but in general principles. Those principles, even as we begin to grasp them, have a way of slipping away from us.

One old dogma that reef biologists have been rethinking has to do with the relative stability of the coral ecosystem. Conventional wisdom taught, until recently, that the simple ecosystems of high latitudes—the community of the Arctic tundra, for example—had boom-and-bust economies. They were prone to wild fluctuations in population. Their lack of diversity led to instability. In contrast, the complex ecosystems of low latitudes—the coral reef and tropical rain forest—were thought to be steady-state systems, with populations in equilibrium. New information suggests that these ecosystems are much more in flux than was supposed.

Plagues of sea stars have helped change our views on the reef's stability. The crown-of-thorns sea star, *Acanthaster planci,* is a coral eater normally found in very low concentrations on Pacific reefs. In the 1960s, its populations exploded, destroying large areas of reef in Australia, New Guinea, the Philippines, Palau, Guam, Samoa, Tahiti, and other Pacific islands. It was assumed at first that some human interference led to the outbreaks—and human agency may indeed figure in them. But evidence suggests that humans are not the sole culprits. A common supposition among experts is that the outbreaks are natural events whose frequency is increasing because of some yet unidentified perturbation.

The reef, the lesson of *Acanthaster* appears to suggest, is a more unstable yet less fragile place than we once thought.

"What we see from the crown-of-thorns, and from cyclone damage, is that you can get major changes of communities and relative abundances in very short periods," says Richard Kenchington, a biologist with Australia's Great Barrier Reef Marine Park Authority. "People will say, 'Oh, that reef's stuffed.' You go out seven or eight years later, and you're sure you're at the wrong reef. Because you've got lots of pretty corals."

The resilience of coral communities is being severely tested in all the warm seas. One of the ironies of coral science—an irony echoed in the tropical rain forest just inland of the reef—is that as fast as we discover new species and wonders, those species and wonders are being erased. Siltation from deforestation and dredging, exploitation for construction materials, uncontrolled collecting, dynamite-fishing, poison-fishing, and just plain overfishing are all devastating reefs worldwide.

Modern coral-reef biology began in 1928, when Dr. Charles Maurice Yonge led the first major biological *(Continued on page 226)*

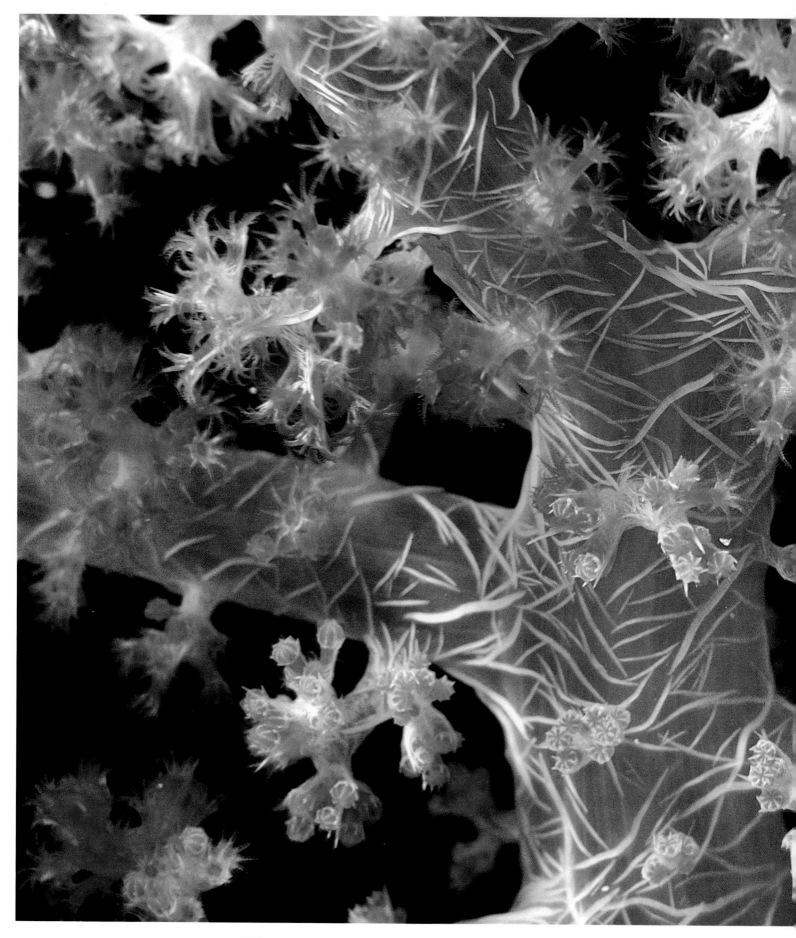

Soft coral polyps extend minuscule tentacles in the Red Sea to snag plankton for nourishment. Coral begins life as a tiny, free-swimming being called a planula, which eventually attaches to a firm surface and becomes a polyp. The polyp extracts calcium and carbonate ions from seawater and manufactures a limestone cup to shelter its soft body. The polyp multiplies and, as each individual adds its stony secretions, coral forms. If the conditions are right, coral colonies grow into coral reefs.

The five-foot-long Caribbean reef shark, also known as the sleeping shark, dwells in shallow coastal waters of the tropics. It is often spotted entering caves, such as this one near Isla Mujeres, Mexico, where it may lie motionless for several hours at a stretch, appearing to sleep. A school of bar jacks hovers above.

225

Male manatees in Florida waters pursue a female, her back covered with algae. These sluggish vegetarians each eat up to 200 pounds of aquatic plants a day, 5 to 10 percent of their body weight. Mangrove forests (opposite) are also found along the Florida coast, as well as in other subtropical and tropical regions worldwide, especially where rivers meet the sea. Their roots trap soil that would otherwise be washed away.

expedition to the Great Barrier Reef. The site of the expedition's studies was Low Isles, east of Port Douglas on the Queensland coast. Low Isles make the most minimal of archipelagoes; just two islands, the one a small, sunny cay of coral sand, with lighthouse, the other a dark wilderness of mangroves. The little two-isle archipelago has a nice polarity. It represents beautifully the two great companion ecosystems of shallow tropical seas: the coral reef and the mangrove forest.

Yonge and his team of 18 scientists spent nearly 13 months camped on the sand cay beneath the lighthouse. On trips to the surrounding reefs, they did their pioneering work. Darwin before them had deduced, brilliantly and correctly, how coral atolls form, and other biologists had dabbled in corals, but Yonge's was the first intensive, cross-disciplinary investigation of the workings of the reef. Yonge and his colleagues were repelled by the mangroves and avoided them.

"Passing further inward," Yonge wrote of the mangrove swamp,

"avoiding the innumerable snags which continually obtruded themselves at every possible angle . . . a darker, dismal region, which forbade further passage except to the most adventurous, was gained. This was a region of mud, of black and decaying stumps and dead trees that lay procumbent like long-extinct reptiles wallowing in a steaming Jurassic swamp."

This is how much of the world viewed mangrove forests in those days: as gloomy, vaguely pestilential places. This is how much of the world sees them still. If science came late to the coral reef, it came even later to the mangroves. Serious study did not begin until the 1960s. The mangroves are a huge, maligned, and relatively unknown province of life.

Mangroves are halophytes, "salt-lovers." They are trees wonderfully adapted to their transitional and ambiguous existence between land and sea. Mangrove leaves have waxy cuticles, protection against excessive transpiration of water, which is the principal stress of that hot, salty environment. The mud of mangrove swamps is poor in oxygen, and the trees have made a number of adaptations to that. In species like the red mangrove, thickets of aerial roots—prop roots—buttress the main trunk, gathering nutrients, collecting debris, and respiring for the tree.

Mangrove seeds germinate on the branch. From each fruit grows a single six- to twelve-inch, green, pointy seedling. Dropping at low tide, the successful seedling buries itself like an arrow in the mangrove mud. Taking root, it grows up in the shadow of its parent. Dropping at high tide, the seedling is carried off by the sea. For a while it floats horizontally at the surface; then the tip becomes waterlogged and drops, the seedling shifting to vertical. Somewhere far from the parent tree, a falling tide lowers it gently, and the tip pierces the mud of new territory.

Mangroves are commonest in the tropics, where they cover about 25 percent of the coastline. They are to tropical shores what salt marshes are to temperate. Like the salt marshes of the north, mangroves are important nurseries for sea life. Hatchling and juvenile fish and crustaceans find sustenance in the mangrove leaf litter and mud. The trees have proved capable of shedding more than three tons of leaves per acre per year. The leaves are broken down and consumed by fungi and bacteria, which in turn are eaten by microscopic worms and crustaceans, which are eaten by small fish, which are eaten by larger fish and birds. Life wells out from the mangroves into surrounding seas, benefitting large traditional and commercial fisheries. The mangroves shelter numerous coral-reef creatures in various stages of their life cycles. Mangrove trees provide firewood and timber for coastal people. They serve as natural sewage-treatment plants. They protect coastlines against storms and erosion. They are vitally important to tropical economy, both human and natural.

They are disappearing fast. Mangrove forests have been reduced by 75 percent in Puerto Rico, by 20 percent in peninsular Malaysia. Between 1967 and 1976, almost half the mangrove forests in the Philippines disappeared. Defoliants sprayed during the Vietnam War killed most of the mangrove in the Mekong Delta. Statistics from almost everywhere else in the tropics are similarly disheartening.

We need a new aesthetic for the mangrove forest, one to replace the apprehensive view of Dr. Yonge and the other post-Victorians. Recently the jungle has become the rain forest, among enlightened folk, and something similar needs to happen with the mangrove swamp.

Some sort of transliteration is necessary, often, if one is to grasp the extraordinary richness of tropical marine systems.

It was that way for Tony DeBrum, a legislator in the Republic of the Marshall Islands. DeBrum knew all about the coral reef, having grown up on the Marshallese atoll of Likiep. It was not until he was a student at the University of Hawaii, though, that the reef's richness came home to him. In his student days, DeBrum earned spending money as a lexicographer and he became a co-author of a Marshallese-English dictionary. After many months of compiling Marshallese terms, DeBrum and his colleagues thought to test their retrieval system. They instructed the computer to find all Marshallese words with "fish" or "fishing" in them. It was a diabolical request, though they asked in all innocence. For a long time, the computer sat silent, stunned by its task, or lost in calculations.

"Then it went crazy," DeBrum remembers. "It began printing, and it didn't stop."

Great scrolls of computer paper marched across the floor. When the computer had finished all the names of fish, it drew a breath and began a list of fishing terms. The coral reef is one of the most diverse and complex ecosystems known, and it makes for diversity in the technology of the folk dependent on it. In Marshallese there are terms for: pole fishing for goatfish; pole fishing from the beach; pole fishing from a raised platform or tripod used for fishing; pole fishing from the reef edge at low tide on dark nights; fishing for squirrelfish in small holes on the reef during low tide; hanging onto the reef while spearing; inserting the bare hand into holes in the coral for the fish dwelling there; striking needlefish with a paddle as they float at the surface on moonlit nights; waiting with a spear along the usual paths of fish across the reef; and so on, and so on, for page after page. There is a Marshallese term for line fishing at night by jerking the line to stir bioluminescence. There is a term for fishing the lagoon with hermit crabs as bait. There is a term for hunting porpoises and pilot whales in the lagoon with stones.

"The method is called *jibke*," says DeBrum of this stone hunting, once

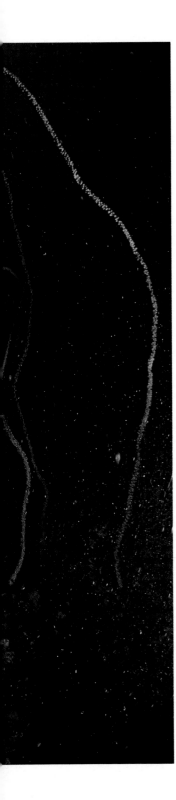

much practiced, seldom now. "The men bang the rocks together underwater to herd the whales. The rocks are shaped to fit the hands—kidney shaped. In Marshallese the word for stone is *deka*. Those special stones are called *deka in jibke*, and that is also our term for kidneys."

An odd underwater clicking reaches the ears of the pilot whales in Ailinglapalap lagoon. The curious juvenile, breaking off its investigation of a coral head, swims up to the surface. Spy-hopping to investigate, the whale sees a skirmish line of paddling canoes advancing.

There are ten canoes, with two or three Marshallese paddlers in each one. Now and again their chief points downward with his paddle, and everyone dives overboard. Five or six feet underwater, they knock their kidney stones together. Some of the whales breach, which allows the Marshallese to see where the pod is moving. The men paddle hard in that direction, tightening the noose about the whales.

The pilots are being hunted with sound—a dose of their own medicine. The line of canoes cuts them off from the pass, trapping them against the inner curve of the atoll's reef. The Marshallese men, jumping overboard repeatedly to knock their rocks together, draw the noose ever tighter, herding the whales closer and closer to the lagoon beach. When the whale that the Marshallese call the "king" or "queen" jumps up on the sand, they have discovered, then all the others will jump after it. Whale meat tastes wonderful, the older islanders say, and it's a cure for constipation. The beached whales will provide, in addition to their flesh, a nice lesson in the dangers of blind obedience to leaders.

Suddenly, as if on signal, the pilots dive under the line of canoes. The loud clicking of kidney stones has brought the dominant bull to his senses, perhaps, or the pod has rebelled against his leadership. For some reason as mysterious as the reason they came in, the whales leave. As the whales pass in waves beneath the boats, the juveniles roll on their sides to look up at the scarred wooden undersides of the hulls.

The pod streams out through the pass. A curious juvenile detours to glide along the reef's outer wall. It makes a small course correction to avoid a whip coral corkscrewing out 15 feet from the wall. The tiny, translucent goby inhabiting the whip darts around, quick as a squirrel, to the opposite side of the stalk. Several juvenile Gaimard's wrasses stare back at the juvenile pilot. The wrasses, at this stage of life, are a bright orange marked in odd places with great daubs of white.

Sociability is strong in pilot whales and it quickly overcomes curiosity. The young whale, veering from the hallucination of the coral, hastens to catch up with the others. Smiling fixedly, its flukes pumping, it sails homeward into the vibrant and featureless blue of the open sea.

Only 10 to 15 million years old, Caribbean coral reefs are less diverse than their 60-to-70-million-year-old Pacific Ocean counterparts, but they are rich in other marine life. Near Grand Cayman Island, divers attract stingrays with squid and other invertebrates.

Stingrays locate food using highly developed electroreceptors and finely tuned senses of smell and touch. They suck prey into their mouths and crush it with grinding plates.

The stingray often lies partially buried in the sea's sandy bottom (below). Its whiplike tail can drive a tail spine into an intruder, inflicting a painful wound.

Pages 234–235: With undulating movements, two

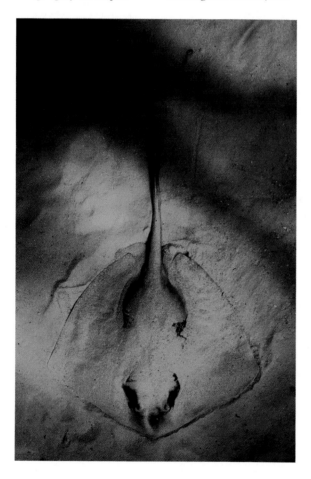

Above, Right, and Pages 234–235: *Dasyatis americana*, Southern Stingray, 4 feet long

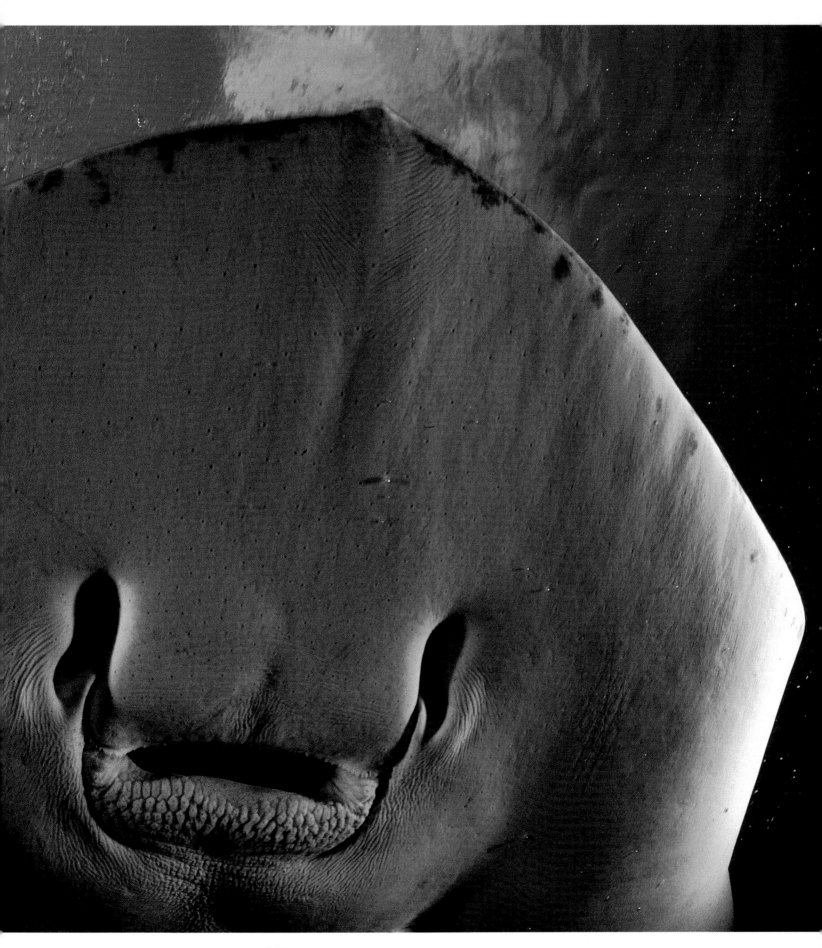

233

Agelas sp., Tube Sponge

Scarus vetula, Queen Parrotfish, up to 20 inches long

rays soar over the bottom sands of the Caymans' North Sound.

Tube sponges (left) also thrive in the Caymans' tropical seas, as do brown, orange, and purple trumpet sponges. Reaching nine feet in length, tube sponges grow on the slopes of coral reefs. They take in water through pores and expel it through their tubes.

Sponges are the most primitive of multicelled animals. They have no nervous, digestive, or circulatory systems—in fact, no real systems at all. Sponges consist of only a few kinds of cells arranged in layers; each kind seems to act independently of the others. For support, many sponges possess needlelike internal structures called spicules.

A queen parrotfish (above) off George Town harbor in Grand Cayman, largest of the three Cayman Islands, blows a protective bag around itself at night from a gland in its mouth. The bag may mask the fish's scent from predators. Or maybe the sand-flecked mucous cocoon acts as a tripwire to wake

the sleeping parrotfish if danger approaches.

A school of juvenile crevalle jacks routs a great barracuda (right) only moments after it had attacked them. Closing ranks, the jacks forced the larger barracuda to retreat. Although fish show no expression, these jacks look outraged.

Crevalle jacks are said to rub against larger fish to rid themselves of parasites. Found throughout the Caribbean and other warm Atlantic waters, they also inhabit the Pacific from the Gulf of California to Peru, near Hawaii, and along the coasts of Asia.

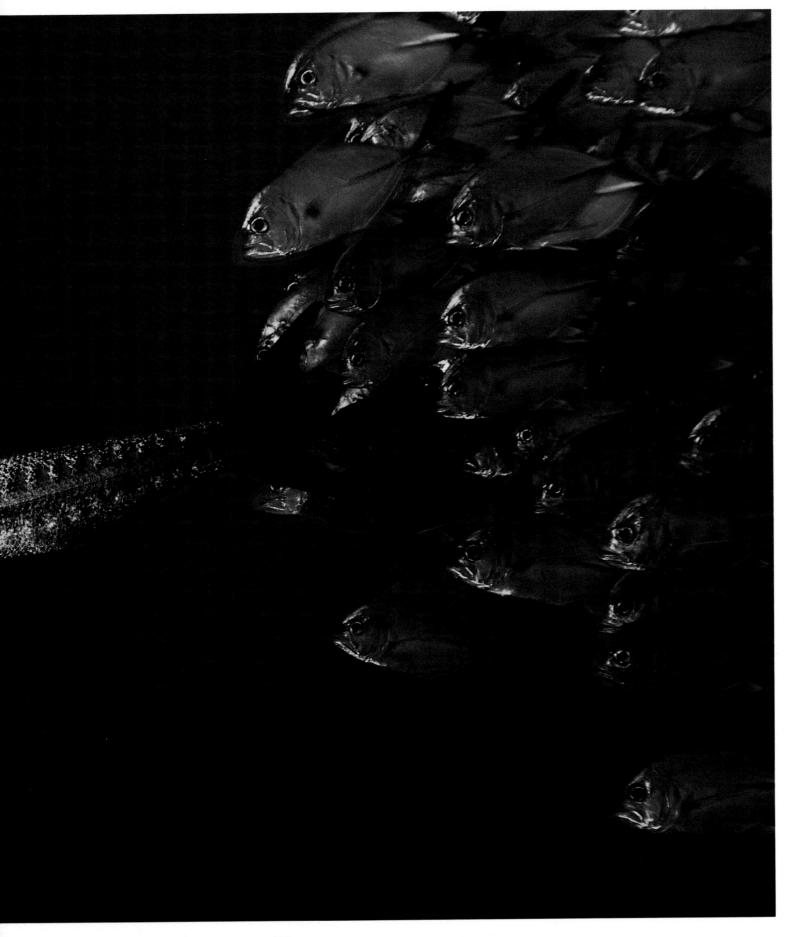

Mugil sp., Mullet, 18 inches long

The Kuroshio (Japan Current) brings warm waters and tropical marine life to Japan's seas. Translated as "black stream," the Kuroshio supports fertile fishing grounds in Suruga Bay and neighboring Sagami Bay. Both are off the south coast of Honshu, Japan's largest island.

Four large mullet (left) swim amid submerged rocks of Izu Oceanic Park, a corner of Sagami Bay. Izu is part of Japan's extensive system of marine parks, controlled jointly by a national park service and commercial fishermen. The torpedo-shaped mullet sometimes escape fishermen's nets by leaping several feet out of the water and over the sides.

As dusk descends on Mount Fuji, two squid boats on Suruga Bay (below) shine powerful floodlights to attract the delicacy known to Japanese as *ika.* About 120 species of squid, cuttlefish, and other cephalopods live in Japanese waters. One, the *hotaruika,* or fire squid, has been designated a national monument because of its luminescence.

Pages 242–243: A school of cardinalfish threads its way through Suruga Bay's forest of wire coral, a form of black coral. Cardinalfish typically dwell in coral reefs or rocky areas where hiding places abound, coming out to feed only at night. In some species, the male guards the eggs by carrying them in his mouth.

A squid, such as *Sepio-teuthis,* has eyes almost as

Pages 242–243: *Apogon* sp., Cardinalfish, 3 inches long; *Cirrhi-pathes* sp., Wire Coral

Sepioteuthis lessoniana, Bigfin Reef Squid, 2 feet long

complex as a human's (above). Two of the squid's ten tentacles have evolved into long arms with four rows of suckers for catching prey. In some areas, squid gather in the thousands to mate and lay eggs. Their frenzied nuptials churn the sea like boiling water, with the males repeatedly changing color. Females anchor each egg capsule to prevent its washing away.

A scorpionfish (right) can alter its color and texture to match its surroundings. It often lies hidden on the seafloor waiting quietly for an unwary victim to swim by. Scorpionfish possess strong, sometimes venomous fin spines that can inflict deep, painful wounds.

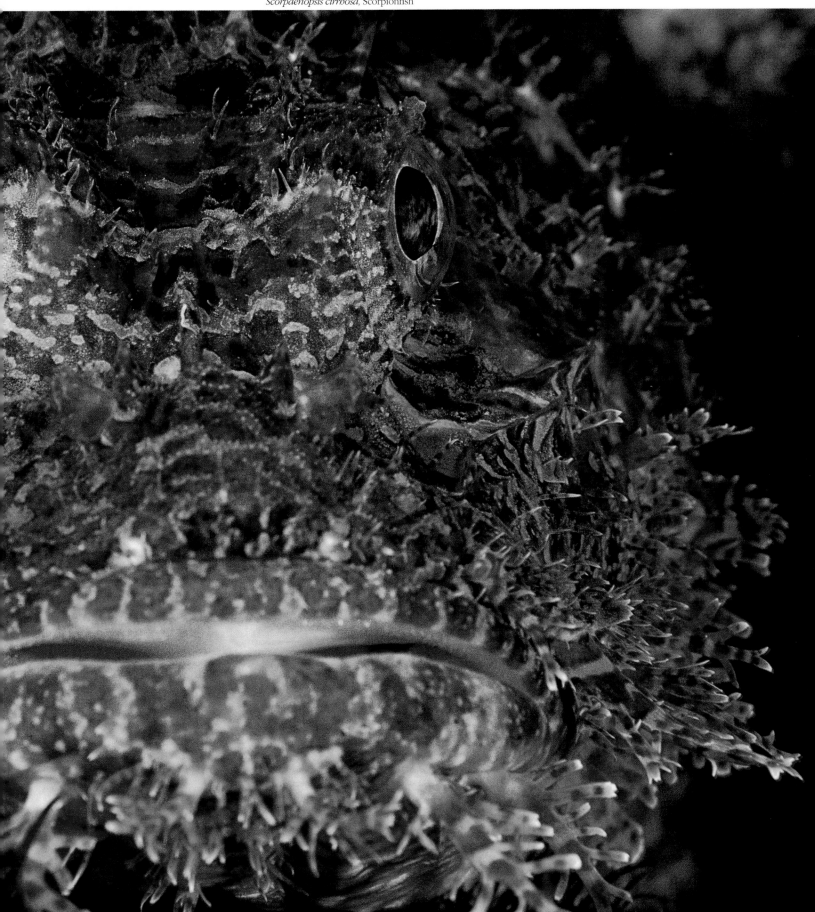

Cerianthus filiformis, Tube Anemone, about 10 inches across

The flowerlike cerianthid (above) lacks the pedal disc, or foot, characteristic of most sea anemones. Instead, cerianthids build tubes composed of sand, secreted mucus, and the same kind of stinging cells found in their tentacles. Half buried in the sea bottom, they withdraw into their tubes when disturbed. Like other anemones, cerianthids use their long, slender tentacles to capture prey.

With their brilliant red color, the male cherry-blossom anthias (right) advertise their presence to the drabber females. Each male's white spots differ slightly. If the male population declines, mature females can change sex to make up the difference.

Sacura margaritacea, Cherry-blossom Anthias, up to 5 inches long

Muraena pardalis, Harlequin Moray Eel, 2.5 feet long

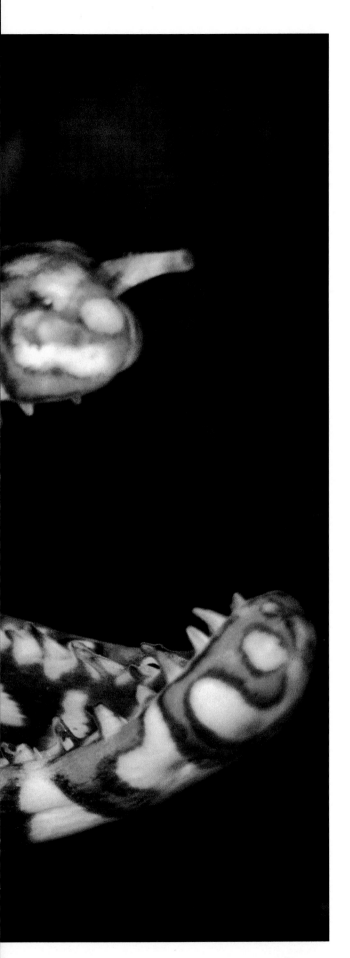

Apterichtus moseri, Finless Snake Eel, 1 foot long

The harlequin moray eel (left), like other morays, lurks in holes or between boulders. Basically shy, it only attacks when threatened or hungry. It usually emerges at night, and can swallow a fish or a crab in one gulp. The harlequin rarely strikes divers, but can mistake human hands for prey, sometimes inflicting nasty wounds. Hornlike nostrils protrude from the top and front of its snout. Called "tiger eel" by the Japanese, the harlequin's bright stripes continue inside its ferocious jaw.

The open-mouth gape of the slender snake eel (above) aids in breathing. Its spikelike tail helps it burrow into the sand, and swim backward.

The Red Sea stretches 1,300 miles from Bab al Mandab, the strait at its southern end, to the Sinai Peninsula and the Gulfs of Suez and Aqaba in the north. Only 190 miles at its widest, the Red Sea was formed some 25 million years ago as Africa and the Arabian Peninsula drifted apart.

Though no one knows for sure how the Red Sea got its name, one theory holds that it came from certain algae, *Trichodesmium erythraeum,* which turn the normally intense blue-green water a reddish brown when they die. The sea itself contains the ocean's warmest and saltiest waters. A continental shelf gives way to waters reaching depths of more than 7,000 feet.

The diversity and beauty of underwater life in the Red Sea stand in stark contrast to the gnarled, barren rock of Râs Muhammad (below). Located at Sinai's southern tip and connected to it by a narrow land bridge, Râs Muhammad (Arabic for "head of Muhammad") has fossil coral cliffs that overlook a creek and a mangrove stand.

Right: *Anthias squamipinnis,* Scalefin Anthias, 4 inches long

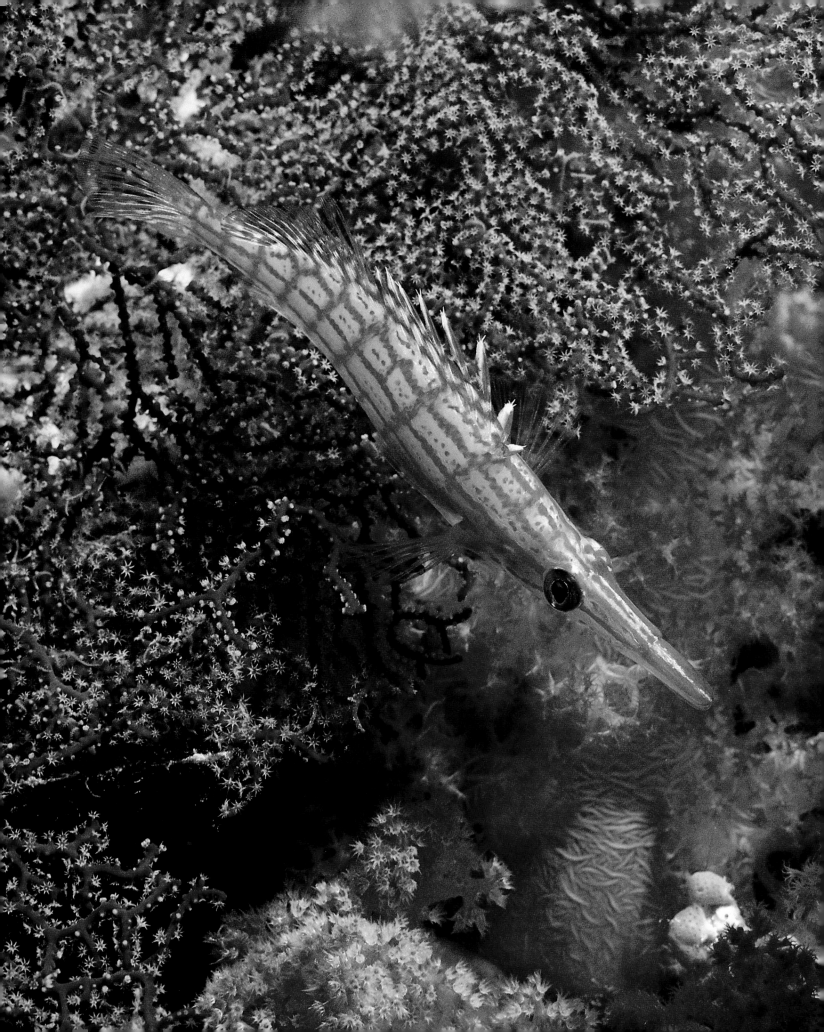

Opposite: *Oxycirrhites typus,* Longnose Hawkfish, 4 inches long

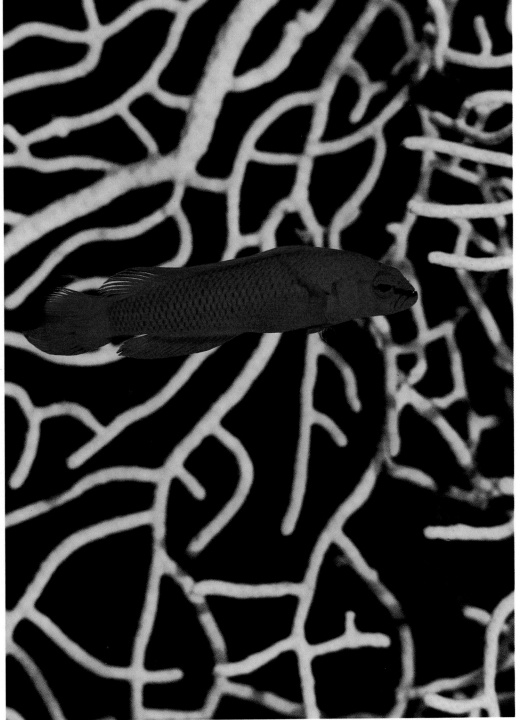

Pseudochromis fridmani, Orchid Dottyback, 2 inches long

Egypt has declared the coral reefs offshore a national marine park.

A school of brightly colored anthias (pages 250-251) swarms over coral in search of food. Virtually all of the fish's eggs hatch into females. As they mature, some change sex to male. If one or more males die, the same number of the largest females will change into males.

In a typical pose, the longnose hawkfish (opposite) rests on its elongated pectoral fins amid gorgonian coral. The hawkfish uses its long snout to nab small crustaceans. The orchid dottyback (left), here seen in gorgonian fan coral, is usually found on vertical rock faces or beneath overhangs.

Hipposcarus harid, Longnose Parrotfish, up to 30 inches

Large schools of long-nose parrotfish (above) gather at Râs Muhammad to mate and lay eggs. Parrotfish are named for their bright colors and beaklike fused teeth. They feed on algae growing on lime-stone or coral, using their strong jaws to grind the rock and extract food. The discarded material adds to the sand below.

The humphead wrasse (right) is as imposing a sight on a coral reef as in this photo. The largest wrasses known to science, humpheads can weigh 400 pounds or more. Adult males develop a promi-nent hump on their fore-heads starting just before the eyes. Humpheads live in relatively shallow waters at the reef's edge.

Cheilinus undulatus, Hump-head Wrasse, 6 feet long

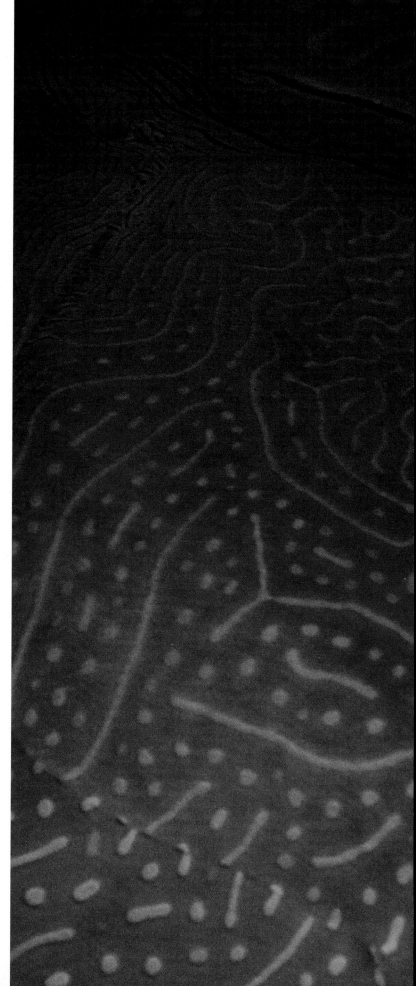

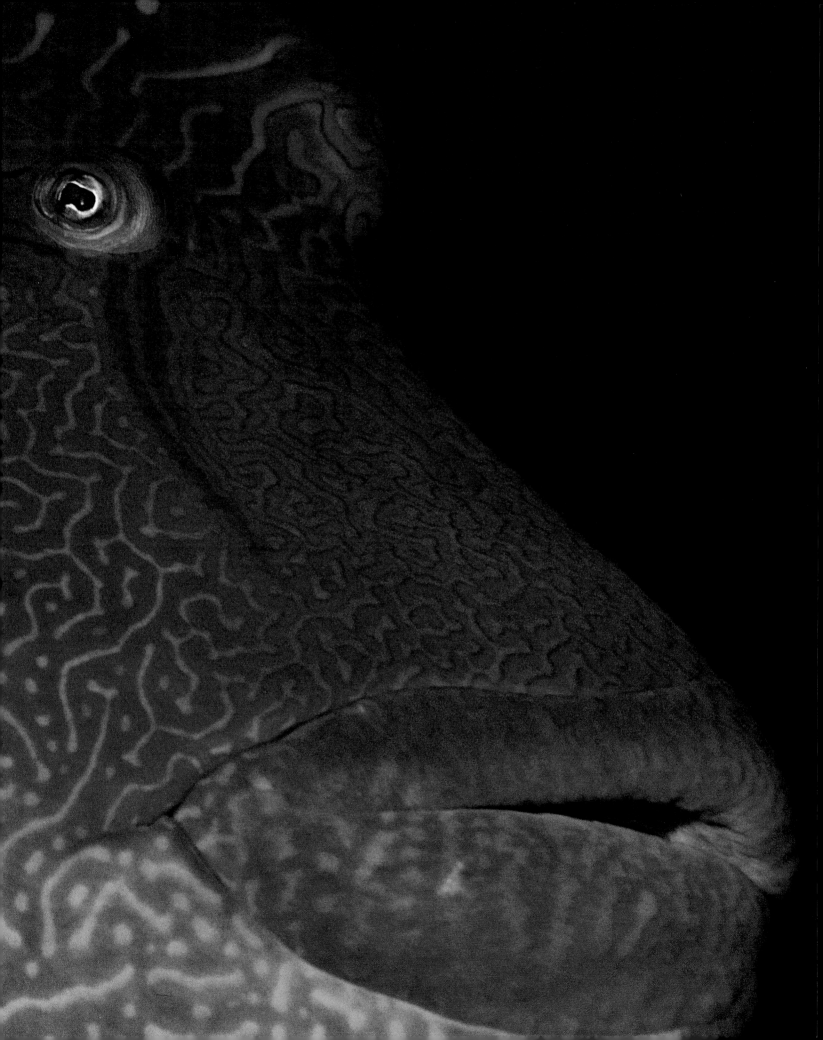

A sign of the times, a half dozen gray moray eels (left) emerge from a discarded oil drum lying in the sand in the Gulf of Aqaba. The most common Red Sea morays, these almost white eels are more gregarious than their kin elsewhere. Like other morays, however, they lack pectoral fins and scales.

Pages 258–259: In the shallow, sandy bottoms of the Red Sea, colonies of serpentine garden eels dance back and forth in the current. Often only six inches apart, hundreds or even thousands of individual eels can populate a single colony. Looking like grass blades blowing in the wind, they curtsy, bow, and sway as if spellbound by a snake charmer.

Garden eels dig burrows in the sand. They moor themselves in the mucus-lined burrows by their tail tips, extending most of their three-foot-long bodies out in search of plankton. If they sense danger, the eels quickly retreat into their burrows, even swimming through the sand to escape.

Garden eels' mating dances begin when an olive-spotted male reaches toward a female while rippling his white dorsal fin. At times, the male may wave his head from side to side and sway about his mate. The female then entwines her slightly shorter body around the male. The dance may be repeated a dozen times in a single morning.

Pages 258–259: *Gorgasia* sp., Garden Eel

The country of Papua New Guinea occupies the eastern half of New Guinea, the world's second largest island, lying off the north coast of Australia. Papua New Guinea includes island chains to the north and east.

Just south of the Equator, Papua New Guinea's tropical seas harbor some of the world's most extensive and diverse coral reefs. The most profuse reefs are found on steep slopes and nearly vertical flanks of outlying islands.

Although Papua New Guinea's coral reefs are virtually pristine compared to those of many other countries, they too are threatened by pollution. Deforestation and poor land management, in particular, are increasing soil erosion into surrounding seas.

Young Papua New Guineans (opposite) paddle their outrigger canoe over billowing sea fans growing in Milne Bay off New Guinea's eastern tip. Waving with the flowing current, sea fans and other gorgonian corals dance in the shade of trees on an overhanging bank.

The seas off Cape Vogel (below) contain some of Papua New Guinea's richest coral reefs. Fed by natural land runoff from abundant rainfall, the reefs cling to steep slopes offshore. The Solomon Sea north of Cape Vogel is characterized by deep trenches, some reaching nearly 30,000 feet.

Opposite: *Melithaea* sp., Gorgonian Coral

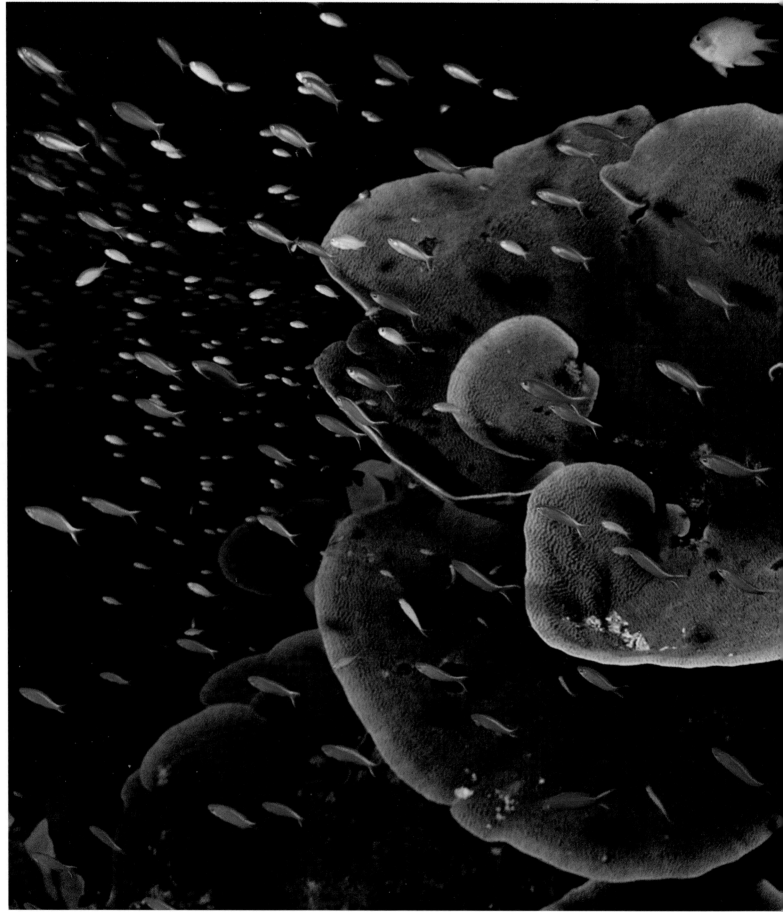

Pomacanthus xanthometopon, Yellow-faced Angelfish, 1 foot long

Like a living rainbow, purple anthias and yellow damselfish (left) stream past green-and-yellow hued chalice coral near Milne Bay. Damselfish, few of which measure over six inches long, favor shallow tropical and subtropical waters. Some live safely among sea anemones by coating themselves with anemone mucus, so that the anemones cannot distinguish the damselfish from themselves.

Baring its sharp teeth, a yellow-faced angelfish (above), off New Ireland Island, uses its vivid yellow and blue coloration to defend its territory. Trespassers are warned away by a twisting display. The angelfish grazes sponges in the coral shallows.

A school of barracuda encircles a lone diver near New Hanover Island. Curious creatures, barracuda will follow swimmers, boats, and even people walking along the shore. The torpedo-shaped great barracuda lead solitary lives as they grow older and larger. Some 20 species are found worldwide, mostly in tropical seas.

Sphyraena jello, Pickhandle Barracuda, 3 feet long

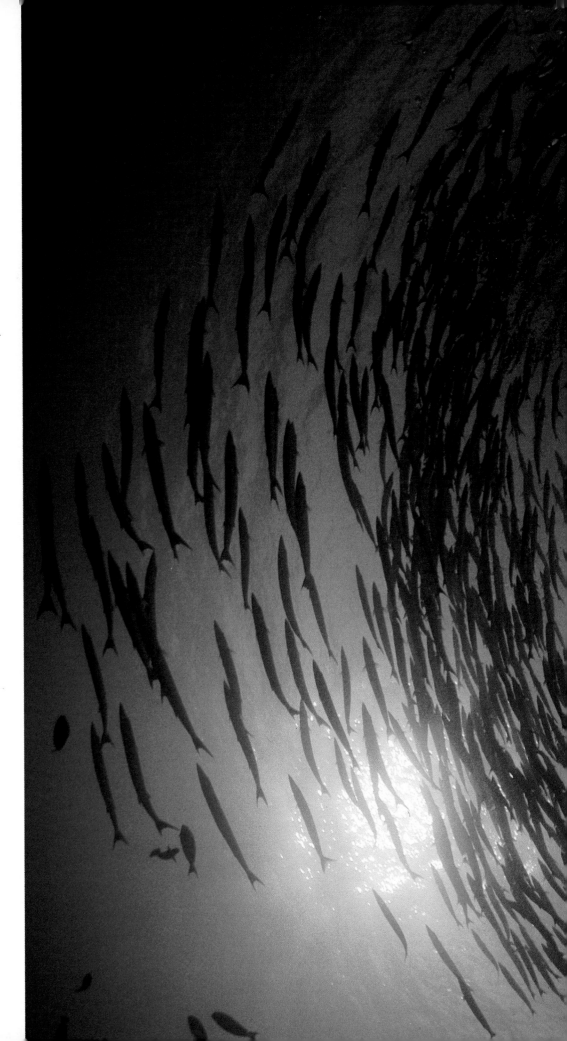

Stretching more than 1,250 miles south from Australia's northeastern tip, the Great Barrier Reef is the largest structure on earth built by living organisms. The reef consists of thousands of individual reefs, shoals, and islets, many dry or barely awash at low tide.

The Great Barrier Reef was created by corals, tiny organisms that often live in huge colonies. In all, at least 350 species of coral populate the reef. They provide shelter and food for countless other marine creatures. Indeed, corals constitute only about 10 percent of the reef's life.

Ribbon Reef Number 10 (below) is the longest of the Great Barrier's outer reefs. Ribbon reefs form elongated reef slopes at right angles to prevailing winds. Typically, such reefs sweep in graceful crescents.

Among the reef's most common fish, fairy basslets (right) swim through pretzel-like gorgonian coral. All fairy basslets hatch into females, some changing into males like some species of anthias. If a male dies, the dominant

Right: *Pseudanthias tuka*, Fairy Basslet, 4 inches long

Oxymonacanthus longirostris, Beaked Leatherjacket

female changes sex and replaces him.

Three-to-four-inch-long beaked leatherjackets (above) eat small crustaceans and coral polyps, the living coral animals. The leatherjackets, which here dart up and down among antler coral branches, prefer the Great Barrier's outer reef edges.

Looking like delicate plates and tables, fast-growing *Acropora* corals (right) bask under the reef's breaking waves. The leaflike plates are sometimes arranged in stacks. The dominant corals in many parts of the Great Barrier Reef, *Acropora*—like most corals—are hermaphroditic, releasing both sperm and egg cells. Other *Acropora* species grow from low bushy clumps into large branching colonies resembling stag horns.

Pages 270–271: An uncommon blenny-like fish with three fins hides among coral. The fish's first two fins consist of slender, immobile spines, while the third has at least seven soft rays.

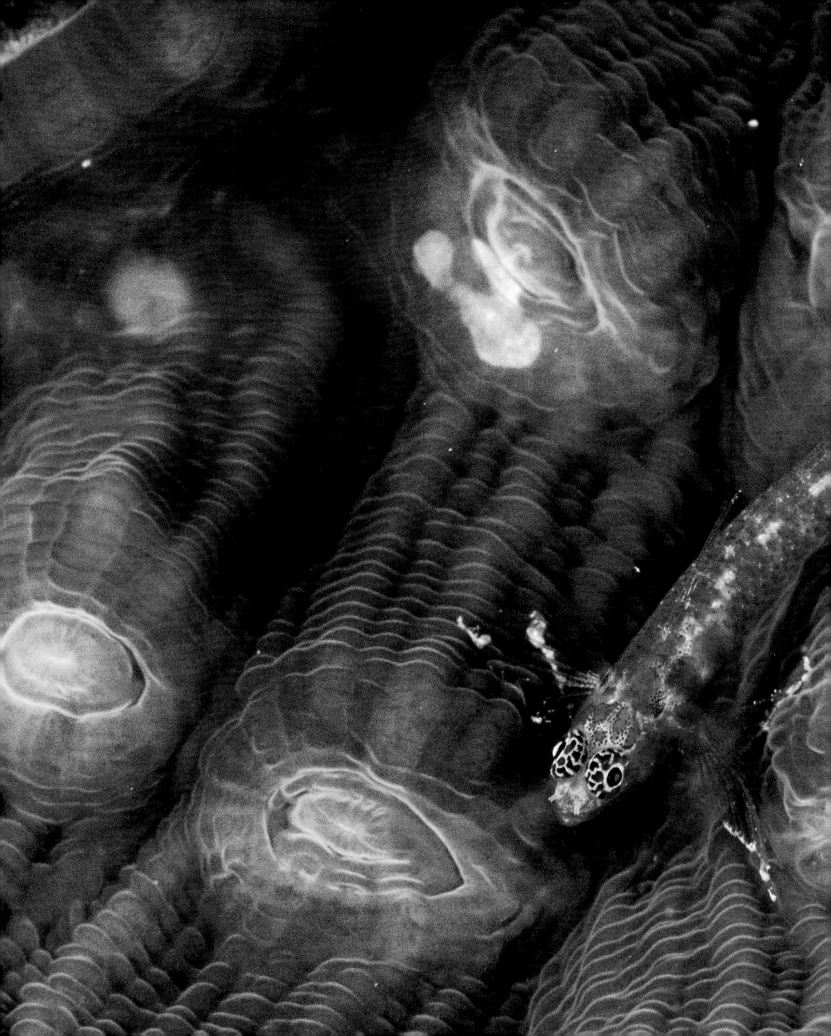

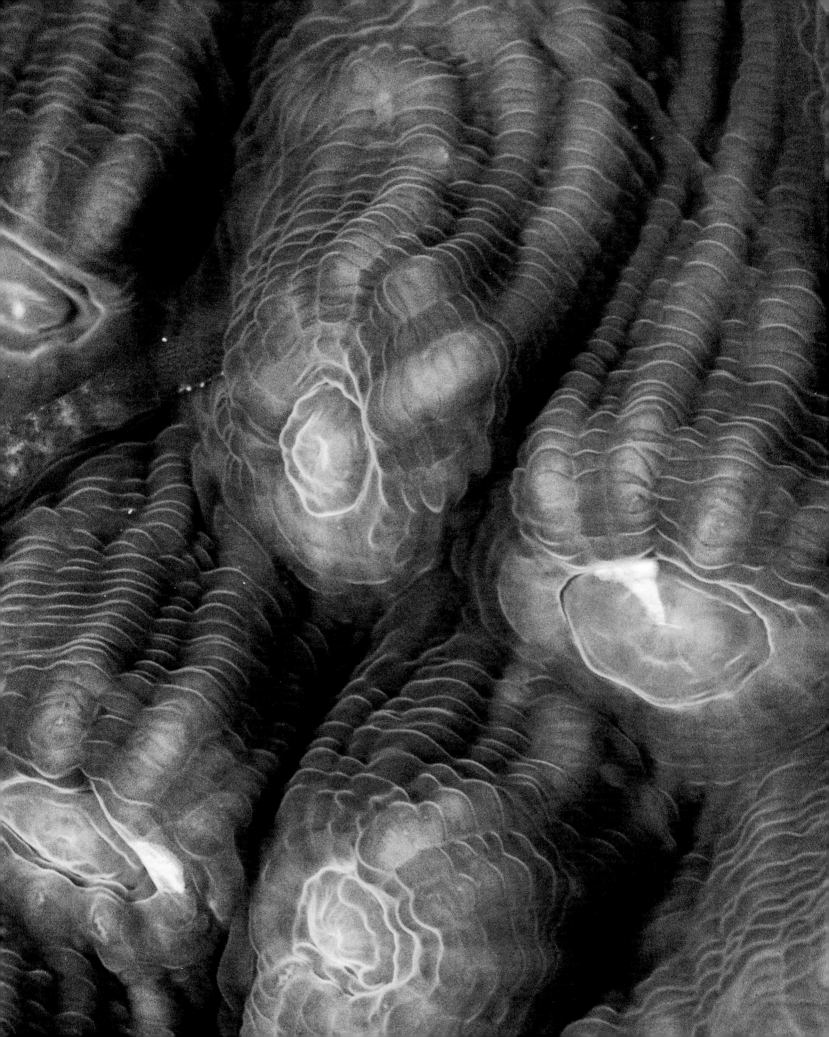

Born in San Francisco in 1944, Kenneth Brower has written about nature, ecology, and wilderness in several books, including *The Starship and the Canoe, Wake of the Whale,* and, for the National Geographic Society, *Yosemite: An American Treasure.* His work has also appeared in numerous magazines, among them *National Geographic, National Geographic Traveler, The Atlantic, Audubon,* and *Smithsonian.* His interest in the ocean has taken him to Australia, Micronesia, and the Galápagos Islands. He lives in Oakland, California, with his wife and two children.

ACKNOWLEDGMENTS

The Book Division is grateful to many individuals and organizations for their assistance during the preparation of *Realms of the Sea.* Special thanks go to David Doubilet, D. Ann Pabst of the University of British Columbia, William A. McLellan, and David A. Ross of the Woods Hole Oceanographic Institution.

The following institutions and their staffs were generous with their time and resources: Harbor Branch Oceanographic Institution, Inc., Fort Pierce, Florida; Monterey Bay Aquarium and its Research Institute, Monterey, California; Scripps Institution of Oceanography, La Jolla, California; Smithsonian Institution, Departments of Invertebrate and Vertebrate Zoology, Washington, D.C.; and Woods Hole Oceanographic Institution, Woods Hole, Massachusetts.

For their help, we wish to thank Frank Aikman III, NOAA/National Ocean Service; Stephen P. Alexander, Wadsworth Center for Laboratories and Research; Stephen D. Cairns, Fenner A. Chace, Jr., Kristian Fauchald, Ian G. Macintyre, Lynne R. Parenti, David L. Pawson, Clyde F. E. Roper, Klaus Ruetzler, and Victor G. Springer, Smithsonian Institution; James J. Childress and Kenneth C. Macdonald, University of California at Santa Barbara; Eugenie Clark, University of Maryland; Bruce B. Collette, NOAA/National Marine Fisheries Service; Sylvia A. Earle, NOAA; Linda Guinee; Robert R. Hessler, Scripps Institution of Oceanography; Terrence Jach; Steven K. Katona, College of the Atlantic; Ron Larson, Harbor Branch Oceanographic Institution, Inc.; Jennifer MacPherson, Center of Marine Biotechnology; Glenn P. Moffat, Foothill College; John E. Pequegnat, Humboldt State University; Harold V. Thurman, Mt. San Antonio College; David J. Wrobel, Monterey Bay Aquarium; and Bernard Zahuranec, Office of Naval Research.

We are also indebted to the National Geographic Library and its Map and News Collections, Administrative Services, Audiovisual Division, Illustrations Library, Messenger Center, Pre-Press/Typographic Division, and Records Library.

The reader may wish to consult the *National Geographic Index* for related articles and books. The following books may also be of interest. Many are intended for nonspecialist audiences:

General *Invertebrate Zoology* (1987) by Robert D. Barnes; *Pacific Marine Fishes,* Vols. 1-8 (1984) by Warren E. Burgess and Herbert R. Axelrod; *The Facts on File Dictionary of Marine Sciences* (1988) by Barbara Charton; *A Field Guide to the Pacific Coast Fishes of North America* (1983) by William N. Eschmeyer and Earl S. Herald; *The Book of Waves* (1989) by Drew Kampion; *Undersea Life* (1985) by Joseph S. Levine; *The Audubon Society Book of Marine Wildlife* (1980) by Les Line; *The Life of Fishes* (1966) by N. B. Marshall; *Introduction to Marine Biology* (1983) by Bayard H. McConnaughey and Robert Zottoli; *A Field Guide to Atlantic Coast Fishes of North America* (1986) by C. Richard Robins and G. Carleton Ray; *Introduction to Oceanography* (1988) by David A. Ross; *An Introduction to the Biology of Marine Life* (1984) by James L. Sumich; *Elements of Marine Ecology* (1980) by R. V. Tait; *Marine Biology* (1984) by Harold V. Thurman and Herbert H. Webber; *Fishes of the World* (1975) by Alwyne Wheeler.

The Ocean World *An Introduction to the World's Oceans* (1989) by Alyn C. Duxbury; *Oceanography* (1985) by M. Grant Gross; *The Ocean World Encyclopedia* (1980), by Donald G. Groves and Lee M. Hunt; *The Oceans: Their Physics, Chemistry, and General Biology* (1942) by Harold U. Sverdrup, Martin W. Johnson, and Richard H. Fleming.

Deep Ocean *The Arcturus Adventure* (1926) by William Beebe; *The Face of the Deep* (1971) by Bruce C. Heezen and Charles D. Hollister; *Deep Oceans* (1971), edited by Peter J. Herring and Malcolm R. Clarke; *Deep-Sea Biology* (1979) by N. B. Marshall; *The Sea: Deep-Sea Biology,* Vol. 8, (1983), edited by Gilbert T. Rowe; *Life in the Sea,* Readings from *Scientific American* (1982).

Polar Seas *Key Environments: Antarctica* (1985), edited by W. N. Bonner and D. W. H. Walton; *Arctic Animals: A Celebration of Survival* (1986) by Fred Bruemmer; *Seabirds* (1983) by Peter Harrison; *Antarctic Wildlife* (1982) by Eric Hosking and Bryan Sage; *Arctic Wildlife* (1984) by Monte Hummel; *Seals of the World* (1983) by Judith E. King; *The Greenpeace Book of Antarctica* (1989) by John May; *Wild Ice: Antarctic Journeys* (1990) by Ron Naveen, et al.; *Wildlife of the Polar Regions* (1981) by G. Carleton Ray and M. G. McCormick-Ray; *The Arctic Ocean* (1982), edited by L. Rey; *Polar Bears* (1988) by Ian Stirling and Dan Guravich; *Animals of the Arctic* (1971) and *Animals of the Antarctic* (1972) by Bernard Stonehouse; *Arctic and Antarctic* (1982) by David Sugden.

Open Ocean *The Biology of Marine Mammals* (1969), edited by Harald T. Andersen; *Biology and Conservation of Sea Turtles: Proceedings of the World Conference on Sea Turtle Conservation,* Washington, D.C., November 26-30, 1979, (1982), edited by Karen A. Bjorndal; *Whales of the World* (1989) by Nigel Bonner; *Wake of the Whale* (1979) by Kenneth Brower; *The Endless Migrations* (1985) by Roger Caras; *The Sharks of North American Waters* (1983) by José I. Castro; *Whales* (1986) by Jacques-Yves Cousteau and Yves Paccalet; *The*

Natural History of Whales and Dolphins (1987) by Peter G. H. Evans; *The Open Sea–Its Natural History: The World of Plankton* (1956) by Alister C. Hardy; *Galápagos: A Terrestrial and Marine Phenomenon* (1988) by Paul Humann; *Galápagos: A Natural History Guide* (1985) by M. H. Jackson; *A Field Guide to the Whales, Porpoises, and Seals of the Gulf of Maine and Eastern Canada* (1983) by Steven K. Katona, Valerie Rough, and David T. Richardson; *The Sierra Club Handbook of Whales and Dolphins* (1983) by Stephen Leatherwood and Randall R. Reeves; *The Greenpeace Book of Dolphins* (1990), edited by John May; *The World's Whales* (1984) by Stanley M. Minasian; *Conserving Sea Turtles* (1983) by Nicholas Mrosovsky; *With the Whales* (1990) by Flip Nicklin and James Darling; *The Year of the Whale* (1969) by Victor B. Scheffer; *Sounding* (1982) by Hank Searls; *Sharks in Question* (1989) by Victor G. Springer and Joy P. Gold; *Galápagos: Discovery on Darwin's Islands* (1988) by David W. Steadman and Steven Zousmer; *The Sargasso Sea* (1975) by John and Mildred Teal.

Temperate Seas *A Naturalist's Seashore Guide* (1979) by Gary J. Brusca and Richard C. Brusca; *Kelp Forests* (1989) by Judith Connor and Charles Baxter; *A Field Guide to the Atlantic Seashore* (1979) by Kenneth L. Gosner; *Pacific Coast Subtidal Marine Invertebrates* (1980) by Daniel W. Gotshall and Laurence L. Laurent; *Guale, the Golden Coast of Georgia* (1974) by Robert Hanie (includes the essay "Living Marsh" by Eugene P. Odum); *Seashore Life of the Northern Pacific Coast* (1983) by Eugene N. Kozloff; *The Amber Forest* (1988) by Ronald H. McPeak, Dale A. Glantz, and Carole R. Shaw; *The Audubon Society Field Guide to North American Seashore Creatures* (1981) by Norman A. Meinkoth; *Between Pacific Tides* (1939) by Edward F. Ricketts, Jack Calvin, and Joel W. Hedgpeth; *Life and Death of the Salt Marsh* (1969) by John and Mildred Teal.

Tropical Seas *Underwater Paradise* (1989) by Robert Boye; *Light in the Sea* (1989) by David Doubilet; *Key Environments: Red Sea* (1987), edited by Alasdair J. Edwards and Stephen M. Head; *The Coral Reef* (1981) by Alan Emery; *A Field Guide to Coral Reefs* (1982) by Eugene Kaplan; *Micronesian Reef Fishes* (1989) by Robert F. Myers; *Red Sea Reef Fishes* (1983) by John E. Randall; *Reader's Digest Book of the Great Barrier Reef* (1984); *The Undersea Predators* (1984) and *The Underwater Wilderness: Life Around the Great Reefs* (1977) by Carl Roessler; *A Natural History of the Coral Reef* (1983) by Charles R. Sheppard; *Coral Reefs of the World,* Vols. 1-3 (1988), edited by Susan M. Wells.

Periodicals that publish stories about the ocean and its creatures include:

Audubon, BioScience, Geographical, International Wildlife, National Geographic, National Wildlife, Natural History, Nature, New Scientist, Ocean Realm, Oceanus, Science, Science News, The Science Teacher, Scientific American, Sea Frontiers, Smithsonian.

Composition for this book by the Typographic section of National Geographic Production Services, Pre-Press Division. Color separations by Lanman Progressive Co., Washington, D. C.; Phototype Color Graphics, Pennsauken, N.J. Printed and bound by R. R. Donnelley & Sons Co., Willard, Ohio. Paper by Mead Paper Co., New York, N.Y. Dust jacket printed by Peake Printers Inc., Cheverly, Md.

Library of Congress CIP Data

Brower, Kenneth, 1944-
   Realms of the sea / by Kenneth Brower ; prepared by the National Geographic Book Division.
     p.   cm.
   Includes index.
     ISBN 0-87044-855-2 (regular ed.).
     — ISBN 0-87044-856-0 (deluxe ed.)
   1. Marine biology. 2. Ocean. I. National Geographic Society (U. S.). Book Division. II. Title.
   QH91.B73   1991   91-11113
   574.92—dc20   CIP